An Introduction
to
Matrix and Finite Element Methods
in
Civil Engineering

An Introduction to Matrix and Finite Element Methods in Civil Engineering

G. N. SMITH

Senior Lecturer in Civil Engineering,
Heriot-Watt University, Edinburgh

APPLIED SCIENCE PUBLISHERS LTD
LONDON

APPLIED SCIENCE PUBLISHERS LTD
RIPPLE ROAD, BARKING, ESSEX, ENGLAND

ISBN 0 85334 502 3

LIBRARY OF CONGRESS CATALOG CARD NUMBER 79-134294

With 68 Illustrations

Printed in Great Britain by Galliard Limited, Great Yarmouth, Norfolk, England.

Preface

THE advent of computers has meant that matrix algebra is now often used in the solution of civil engineering problems. The recent development of the finite element method has been an important contribution to the field of stress analysis. This book is an attempt to present to the reader a text which, without reference to further literature, will enable him to obtain an understanding of these modern techniques.

As matrices and finite elements are often used in stress analysis two chapters containing a summary of this subject have been included.

Whenever possible I have included worked examples. The applied examples have been largely taken from the field of soil mechanics but most are equally applicable to other fields of civil engineering.

G. N. SMITH

Contents

1

Determinants

INTRODUCTION

WHEN four quantities, *a*, *b*, *c* and *d*, are written in the form:

$$\begin{vmatrix} a & b \\ c & d \end{vmatrix}$$

the expression is called a determinant and the convention is that this formation means $ad - bc$. The diagonal from top left to bottom right (the leading diagonal) gives the term *ad* and, from this product, is subtracted the term *bc*, formed from the other diagonal.

Example

Evaluate

$$\begin{vmatrix} 1 & 2 \\ 3 & 4 \end{vmatrix} \quad \text{and} \quad \begin{vmatrix} 1 & 2 \\ 3 & -4 \end{vmatrix}$$

$$\begin{vmatrix} 1 & 2 \\ 3 & 4 \end{vmatrix} = 4 - 6 = -2 \qquad \begin{vmatrix} 1 & 2 \\ 3 & -4 \end{vmatrix} = -4 - 6 = -10$$

When nine quantities are written in the determinant form:

$$\begin{vmatrix} a_1 & b_1 & c_1 \\ a_2 & b_2 & c_2 \\ a_3 & b_3 & c_3 \end{vmatrix}$$

the convention is that this formation means:

$$a_1 \begin{vmatrix} b_2 & c_2 \\ b_3 & c_3 \end{vmatrix} + b_1 \begin{vmatrix} c_2 & a_2 \\ c_3 & a_3 \end{vmatrix} + c_1 \begin{vmatrix} a_2 & b_2 \\ a_3 & b_3 \end{vmatrix}$$

$$= a_1(b_2c_3 - c_2b_3) + b_1(c_2a_3 - a_2c_3) + c_1(a_2b_3 - b_2a_3)$$

Example

Evaluate

$$\begin{vmatrix} 1 & 2 & 3 \\ 4 & 5 & 6 \\ 7 & -8 & 9 \end{vmatrix}$$

$$= 1(5 \times 9 - 6 \times -8) + 2(6 \times 7 - 4 \times 9) + 3(4 \times -8 \times -5 \times 7)$$

$$= (45 + 48) + 2(42 - 36) + 3(-32 - 35)$$

$$= -96$$

THE EVOLUTION OF DETERMINANTS

Let us assume that three variables, x, y and z, are connected by the three equations:

$$a_1 x + b_1 y + c_1 z = 0 \qquad (1)$$

$$a_2 x + b_2 y + c_2 z = 0 \qquad (2)$$

$$a_3 x + b_3 y + c_3 z = 0 \qquad (3)$$

If we consider equations (2) and (3) and eliminate y:

Equation (2) $\times b_3$ gives $a_2 b_3 x + b_2 b_3 y + c_2 b_3 z = 0$

Equation (3) $\times b_2$ gives $a_3 b_2 x + b_2 b_3 y + c_3 b_2 z = 0$

Hence

$$x(a_2 b_3 - a_3 b_2) = z(c_3 b_2 - c_2 b_3)$$

i.e.

$$\frac{x}{c_3 b_2 - c_2 b_3} = \frac{z}{a_2 b_3 - a_3 b_2}$$

Similarly, if we had eliminated z instead of y:

$$\frac{x}{c_3 b_2 - c_2 b_3} = \frac{y}{c_2 a_3 - c_3 a_2}$$

Hence

$$\frac{x}{c_3 b_2 - c_2 b_3} = \frac{y}{c_2 a_3 - c_3 a_2} = \frac{z}{a_2 b_3 - a_3 b_2} = k \text{ (say)}$$

We can now substitute for x, y and z in equation (1) in terms of k:

$$a_1(b_2 c_3 - c_2 b_3)k + b_1(c_2 a_3 - c_3 a_2)k + c_1(a_2 b_3 - a_3 b_2)k = 0$$

i.e.

$$a_1 \begin{vmatrix} b_2 & c_2 \\ b_3 & c_3 \end{vmatrix} + b_1 \begin{vmatrix} c_2 & a_2 \\ c_3 & a_3 \end{vmatrix} + c_1 \begin{vmatrix} a_2 & b_2 \\ a_3 & b_3 \end{vmatrix} = 0$$

which is usually written as:

$$\begin{vmatrix} a_1 & b_1 & c_1 \\ a_2 & b_2 & c_2 \\ a_3 & b_3 & c_3 \end{vmatrix} = 0$$

As explained the expression

$$\begin{vmatrix} b_2 & c_2 \\ b_3 & c_3 \end{vmatrix}$$

is called a determinant. In this case the determinant consists of two horizontal lines, known as rows, and two vertical lines, known as columns. It has two terms when expanded and is therefore referred to as a second order determinant. In the same manner the expression:

$$\begin{vmatrix} a_1 & b_1 & c_1 \\ a_2 & b_2 & c_2 \\ a_3 & b_3 & c_3 \end{vmatrix}$$

consists of three rows and three columns. It has three terms when expanded and is called a third order determinant.

The terms a_1, b_1, c_1, etc. are called the elements of the determinant.

EXPANSION OF A THIRD ORDER DETERMINANT

In stress analysis third order determinants are fairly common and a quick method of expansion is known as the Rule of Sarrus:

Write down the determinant:

$$\text{Det.} = \begin{vmatrix} a_1 & b_1 & c_1 \\ a_2 & b_2 & c_2 \\ a_3 & b_3 & c_3 \end{vmatrix}$$

And repeat the first two columns as the last two:

$$\begin{array}{ccccc} a_1 & b_1 & c_1 & a_1 & b_1 \\ a_2 & b_2 & c_2 & a_2 & b_2 \\ a_3 & b_3 & c_3 & a_3 & b_3 \end{array}$$

$-$ve $-$ve $-$ve +ve +ve +ve

The expansion of the determinant is the sum of the products of the elements in the lines going from top left to bottom right less the sum of the products of the elements in the lines going from top right to bottom left.
i.e.

$$\text{Det.} = a_1b_2c_3 + b_1c_2a_3 + c_1a_2b_3 - c_1b_2a_3 - a_1c_2b_3 - b_1a_2c_3$$

$$= a_1(b_2c_3 - c_2b_3) + b_1(c_2a_3 - a_2c_3) + c_1(a_2b_3 - b_2a_3)$$

$$= a_1\begin{vmatrix} b_2 & c_2 \\ b_3 & c_3 \end{vmatrix} + b_1\begin{vmatrix} c_2 & a_2 \\ c_3 & a_3 \end{vmatrix} + c_1\begin{vmatrix} a_2 & b_2 \\ a_3 & b_3 \end{vmatrix} \text{ as before.}$$

Example

Evaluate

$$\begin{vmatrix} 1 & 2 & 3 \\ 4 & 5 & 6 \\ 7 & -8 & 9 \end{vmatrix}$$

$$\text{Det.} = \begin{array}{ccccc} 1 & 2 & 3 & 1 & 2 \\ 4 & 5 & 6 & 4 & 5 \\ 7 & -8 & 9 & 7 & -8 \end{array}$$

$-$ve +ve

$$= 1 \cdot 5 \cdot 9 + 2 \cdot 6 \cdot 7 + 3 \cdot 4 \cdot -8 - 2 \cdot 4 \cdot 9 - 1 \cdot 6 \cdot -8 - 3 \cdot 5 \cdot 7$$

$$= 45 + 84 - 96 - 72 + 48 - 105 = -96$$

EXPANSION OF DETERMINANTS HIGHER THAN THE THIRD ORDER

Minors

It has been shown that, by the Rule of Sarrus, the third order determinant:

$$\begin{vmatrix} a_1 & b_1 & c_1 \\ a_2 & b_2 & c_2 \\ a_3 & b_3 & c_3 \end{vmatrix}$$

may be expressed as:

$$\text{Det.} = a_1b_2c_3 + b_1c_2a_3 + c_1a_2b_3 - c_1b_2a_3 - a_1c_2b_3 - b_1a_2c_3$$

which could be written as:

$$\text{Det.} = a_1(b_2c_3 - c_2b_3) - a_2(b_1c_3 - c_1b_3) + a_3(b_1c_2 - c_1b_2)$$

Now

$$b_2c_3 - c_2b_3 = \begin{vmatrix} b_2 & c_2 \\ b_3 & c_3 \end{vmatrix} = A_1 \text{ (say)}$$

A_1 is the determinant obtained by ignoring the row and the column in which the element a_1 appears.
And

$$b_1c_3 - c_1b_3 = \begin{vmatrix} b_1 & c_1 \\ b_3 & c_3 \end{vmatrix} = A_2 \text{ (say)}$$

A_2 is the determinant obtained by ignoring the row and the column in which a_2 appears.
And

$$b_1c_2 - c_1b_2 = \begin{vmatrix} b_1 & c_1 \\ b_2 & c_2 \end{vmatrix} = A_3 \text{ (say)}$$

A_3 is the determinant obtained by ignoring the row and the column containing a_3.
Hence

$$\text{Det.} = a_1A_1 - a_2A_2 + a_3A_3$$

where A_1 is known as the minor of a_1 in the original determinant, and A_2 is known as the minor of a_2, etc.

Similarly, by considering the first row, instead of column, a further expression for the determinant is:

$$\begin{aligned}\text{Det.} &= a_1(b_2c_3 - c_2b_3) - b_1(a_2c_3 - c_2a_3) + c_1(a_2b_3 - b_2a_3) \\ &= a_1A_1 - b_1B_1 + c_1C_1\end{aligned}$$

where

$$B_1 = \begin{vmatrix} a_2 & c_2 \\ a_3 & c_3 \end{vmatrix} \quad \text{and} \quad C_1 = \begin{vmatrix} a_2 & b_2 \\ a_3 & b_3 \end{vmatrix},$$

the respective minors of b_1 and c_1.

Cofactors

In the expansion of a third order determinant by minors a negative sign appears in front of the middle term. With large order determinants several terms can be negative. This is allowed for by the use of cofactors.

The cofactor of a particular term in a determinant is simply the respective minor given a positive or a negative sign. The rule to obtain this sign is:

If the sum of the column and the row of the element considered is even then the sign of the cofactor is positive.

If the sum of the column and row of the element considered is odd then the sign is negative.

Example

$$\begin{vmatrix} a_{11} & a_{12} & a_{13} \\ a_{21} & a_{22} & a_{23} \\ a_{31} & a_{32} & a_{33} \end{vmatrix}$$

$$\text{Cofactor of } a_{11} = + \begin{vmatrix} a_{22} & a_{23} \\ a_{32} & a_{33} \end{vmatrix} \quad (1 + 1 \text{ is an even number})$$

$$\text{Cofactor of } a_{32} = - \begin{vmatrix} a_{11} & a_{13} \\ a_{21} & a_{23} \end{vmatrix} \quad (3 + 2 \text{ is an odd number})$$

Example

Evaluate

$$\begin{vmatrix} 1 & 2 & 3 \\ 4 & 5 & 6 \\ 7 & -8 & 9 \end{vmatrix}$$

Consider the first row, i.e.

$$\text{Det.} = a_1A_1 - b_1B_1 + c_1C_1$$

$$= 1\begin{vmatrix} 5 & 6 \\ -8 & 9 \end{vmatrix} - 2\begin{vmatrix} 4 & 6 \\ 7 & 9 \end{vmatrix} + 3\begin{vmatrix} 4 & 5 \\ 7 & -8 \end{vmatrix}$$

$$= (45 + 48) - 2(36 - 42) + 3(-32-35)$$

$$= 93 + 12 - 201$$

$$= -96$$

Alternatively, consider the first column:

$$\text{Det.} = a_1A_1 - a_2A_2 + a_3A_3$$

$$= 1\begin{vmatrix} 5 & 6 \\ -8 & 9 \end{vmatrix} - 4\begin{vmatrix} 2 & 3 \\ -8 & 9 \end{vmatrix} + 7\begin{vmatrix} 2 & 3 \\ 5 & 6 \end{vmatrix}$$

$$= (45 + 48) - 4(18 + 24) + 7(12 - 15)$$

$$= 93 - 168 - 21$$

$$= -96$$

For third order determinants expansion is best carried out by the Rule of Sarrus but, for determinants of higher orders, i.e. with more rows and columns, expansion by cofactors becomes a useful procedure. Expansion by cofactors can take place using any row or column.

Work is reduced if the column or row initially chosen contains zero elements, as shown in the following example.

Example

Expand

$$\begin{vmatrix} 1 & 2 & 3 & 4 \\ 0 & 1 & 2 & 3 \\ 4 & 0 & 1 & 2 \\ 3 & 4 & 0 & 1 \end{vmatrix}$$

Taking the first column the expansion is:

$$\text{Det.} = 1\begin{vmatrix} 1 & 2 & 3 \\ 0 & 1 & 2 \\ 4 & 0 & 1 \end{vmatrix} + 4\begin{vmatrix} 2 & 3 & 4 \\ 1 & 2 & 3 \\ 4 & 0 & 1 \end{vmatrix} - 3\begin{vmatrix} 2 & 3 & 4 \\ 1 & 2 & 3 \\ 0 & 1 & 2 \end{vmatrix}$$

$$= 1 \times 5 + 4 \times 5 - 3 \times 0 = 25$$

Or, taking the first row:

$$\text{Det.} = 1\begin{vmatrix} 1 & 2 & 3 \\ 0 & 1 & 2 \\ 4 & 0 & 1 \end{vmatrix} - 2\begin{vmatrix} 0 & 2 & 3 \\ 4 & 1 & 2 \\ 3 & 0 & 1 \end{vmatrix} + 3\begin{vmatrix} 0 & 1 & 3 \\ 4 & 0 & 2 \\ 3 & 4 & 1 \end{vmatrix} - 4\begin{vmatrix} 0 & 1 & 2 \\ 4 & 0 & 1 \\ 3 & 4 & 0 \end{vmatrix}$$

$$= 1 \times 5 - 2 \times (-5) + 3 \times 50 - 4 \times 35$$

$$= 25$$

METHOD OF ELIMINATION

Although the method of expansion just described can be used to evaluate any determinant it is obvious that it becomes cumbersome when the minors are large as the process must be repeated several times. An alternative is the method of elimination which can be used in computers to evaluate large determinants and is based on the principle of reducing a determinant expression to a triangular form where only the elements on and above the leading diagonal are non-zero. The principle is based on the fact that when a particular row (or column) of a determinant has, either added to it or subtracted from it a scalar multiple of another row (or column) the value of the determinant is unaltered. (See Property (5) at the end of this chapter.) If this rule is continually applied it is possible to reduce the determinant to the form:

$$\begin{vmatrix} a_{11} & a_{12} & a_{13}\ldots a_{1n} \\ 0 & a_{22} & a_{23}\ldots a_{2n} \\ 0 & 0 & a_{33}\ldots a_{3n} \\ \cdot & & \cdot \\ \cdot & & \cdot \\ \cdot & & \cdot \\ 0 & 0 & 0 \ \ldots a_{nn} \end{vmatrix}$$

When expanded the determinant is simply equal to the products of the elements on the leading diagonal, all the other terms in the determinant involving a zero term.

Example

Evaluate

$$|A| = \begin{vmatrix} 1 & 2 & 3 & 4 \\ 0 & 1 & 2 & 3 \\ 4 & 0 & 1 & 2 \\ 3 & 4 & 0 & 1 \end{vmatrix}$$

Put row (3) = row (3) − 4 × row (1)

row (4) = row (4) − 3 × row (1)

$$|A| = \begin{vmatrix} 1 & 2 & 3 & 4 \\ 0 & 1 & 2 & 3 \\ 0 & -8 & -11 & -14 \\ 0 & -2 & -9 & -11 \end{vmatrix}$$

row (3) = row (3) − 4 × row (4)

row (4) = row (4) + 2 × row (2)

$$|A| = \begin{vmatrix} 1 & 2 & 3 & 4 \\ 0 & 1 & 2 & 3 \\ 0 & 0 & 25 & 30 \\ 0 & 0 & -5 & -5 \end{vmatrix}$$

row (4) = row (4) + row (3) ÷ 5

$$|A| = \begin{vmatrix} 1 & 2 & 3 & 4 \\ 0 & 1 & 2 & 3 \\ 0 & 0 & 25 & 30 \\ 0 & 0 & 0 & 1 \end{vmatrix}$$

$$|A| = 1 \times 1 \times 25 \times 1 = 25$$

More than one row can be altered in an operation. If a row has been altered and is then used, in the same operation, to change another row, the altered, not unaltered, values must be used.

PROPERTIES OF DETERMINANTS

(1) If the elements of a row (or column) of a determinant are multiplied by the cofactors of a different row (or column) and added the result is always zero. This important property of determinants is used in matrix inversions and can be illustrated by an example: Consider the determinant

$$\begin{vmatrix} a_{11} & a_{12} & a_{13} \\ a_{21} & a_{22} & a_{23} \\ a_{31} & a_{32} & a_{33} \end{vmatrix}$$

Let us combine the elements of the first row with the cofactors of the second row:

$$-a_{11}\underbrace{\begin{vmatrix} a_{12} & a_{13} \\ a_{32} & a_{33} \end{vmatrix}}_{\text{Cofactor of } a_{21}} + a_{12}\underbrace{\begin{vmatrix} a_{11} & a_{13} \\ a_{31} & a_{33} \end{vmatrix}}_{\text{of } a_{22}} - a_{13}\underbrace{\begin{vmatrix} a_{11} & a_{12} \\ a_{31} & a_{32} \end{vmatrix}}_{\text{of } a_{23}}$$

$$= -a_{11}(a_{12}a_{33} - a_{13}a_{32}) + a_{12}(a_{11}a_{33} - a_{13}a_{31}) - a_{13}(a_{11}a_{32} - a_{12}a_{31})$$

$$= -a_{11}a_{12}a_{33} + a_{11}a_{13}a_{32} + a_{12}a_{11}a_{33} - a_{12}a_{13}a_{31} - a_{13}a_{11}a_{32} + a_{13}a_{12}a_{31}$$

$$= 0$$

The reader might like to verify this property for different rows and columns.

(2) A determinant is equal to its transpose. Transposes are discussed in Chapter 3. Briefly the process of transposition is to simply change each row into a corresponding column.

$$|\mathbf{A}| = |\mathbf{A}^t|$$

The law can be illustrated by a numerical example:

Example

$$|\mathbf{A}| = \begin{vmatrix} 2 & 3 & 4 \\ 6 & 8 & 2 \\ 1 & 3 & 5 \end{vmatrix}$$

Hence

$$|\mathbf{A}^t| = \begin{vmatrix} 2 & 6 & 1 \\ 3 & 8 & 3 \\ 4 & 2 & 5 \end{vmatrix}$$

Applying the rule of Sarrus:

$$|\mathbf{A}| = 2 \times 8 \times 5 + 3 \times 2 \times 1 + 4 \times 6 \times 3 - 3 \times 6 \times 5 - 2 \times 3 \times 2 - 1 \times 8 \times 4 = 24$$

$$|\mathbf{A}^t| = 2 \times 8 \times 5 + 6 \times 3 \times 4 + 1 \times 3 \times 2 - 6 \times 3 \times 5 - 2 \times 3 \times 2 - 1 \times 8 \times 4 = 24$$

(3) If a determinant contains a complete row or column consisting of zero elements then the determinant has the value zero.

Consider

$$|\mathbf{A}| = \begin{vmatrix} 1 & 2 & 0 \\ 2 & 3 & 0 \\ 4 & 5 & 0 \end{vmatrix}$$

Applying the rule of Sarrus:

$$|\mathbf{A}| = 0 \times 2 \times 5 + 2 \times 0 \times 4 + 1 \times 3 \times 0 - 2 \times 2 \times 0 - 1 \times 0 \times 5 - 0 \times 3 \times 4 = 0$$

Similarily it will be seen that

$$\begin{vmatrix} 0 & 0 & 0 \\ 1 & 2 & 3 \\ 3 & 4 & 5 \end{vmatrix}$$

also equals zero.

(4) If, in a determinant, a row (or column) is a scalar multiple of another row (or column) the value of the determinant is zero.

Let

$$|\mathbf{A}| = \begin{vmatrix} a_{11} & a_{12} & a_{13} \\ a_{21} & a_{22} & a_{23} \\ a_{31} & a_{32} & a_{33} \end{vmatrix} = \begin{vmatrix} ka_{21} & ka_{22} & ka_{23} \\ a_{21} & a_{22} & a_{23} \\ a_{31} & a_{32} & a_{33} \end{vmatrix}$$

Expanding, by rule of Sarrus:

$$\begin{aligned}|\mathbf{A}| &= ka_{21}a_{22}a_{33} + ka_{22}a_{23}a_{31} + ka_{23}a_{21}a_{32} \\ &\quad - ka_{22}a_{21}a_{33} - ka_{21}a_{23}a_{32} - ka_{23}a_{22}a_{31} \\ &= k(a_{21}a_{22}a_{33} + a_{22}a_{23}a_{31} + a_{23}a_{21}a_{32} \\ &\quad - a_{22}a_{21}a_{33} - a_{21}a_{23}a_{32} - a_{23}a_{22}a_{31}) \\ &= 0\end{aligned}$$

(5) When a particular row (or column) of a determinant has either added to it or subtracted from it a scalar multiple of another row (or column) the value of the determinant is unaltered.

Consider

$$|\mathbf{A}| = \begin{vmatrix} a_{11} & a_{12} & a_{13} \\ a_{21} & a_{22} & a_{23} \\ a_{31} & a_{32} & a_{33} \end{vmatrix}$$

Consider the second row to be multiplied by k and added to the first row. The determinant becomes:

$$\begin{vmatrix} a_{11} + ka_{21} & a_{12} + ka_{22} & a_{13} + ka_{23} \\ a_{21} & a_{22} & a_{23} \\ a_{31} & a_{32} & a_{33} \end{vmatrix}$$

Expanding by cofactors the expansion of this determinant is

$$\begin{aligned}&(a_{11} + ka_{21})A_1 - (a_{12} + ka_{22})A_2 + (a_{13} + ka_{23})A_3 \\ &\quad = a_{11}A_1 - a_{12}A_2 + a_{13}A_3 + ka_{21}A_1 - ka_{22}A_2 + ka_{23}A_3\end{aligned}$$

The first three terms gave the determinant A whereas the remaining three terms give the expansion of another determinant.

The expression is therefore:

$$A + \begin{vmatrix} ka_{21} & ka_{22} & ka_{23} \\ a_{21} & a_{22} & a_{23} \\ a_{31} & a_{32} & a_{33} \end{vmatrix} = A$$

(As, from Property (4), the second determinant is zero).

2

Vectors

A vector could be defined as a line in space. The length of the line gives a measure of its magnitude and the slope of the line gives its direction.

To be mathematical we say that a vector is a quantity possessing both magnitude and direction. The direction of a three dimensional vector can be defined in a particular coordinate system by its components, say x, y and z on the three axes $0X$, $0Y$ and $0Z$. In stress work these axes are taken as being mutually orthogonal, perpendicular to each other, but the argument holds if they are not orthogonal. The value of the components will vary with the axes chosen, the vector may have components x', y' and z' when referred to axes $0X'$, $0Y'$ and $0Z'$. A two dimensional vector is shown in Figure 2.1.

The vector has two components, x_1 along axis 01 and x_2 along axis 02. In this particular case $x_1 = 4$ and $x_2 = 2$.

The vector can therefore be written in mathematical form as:

$$\mathbf{y} = (x_1, x_2) \quad \text{or} \quad (4, 2)$$

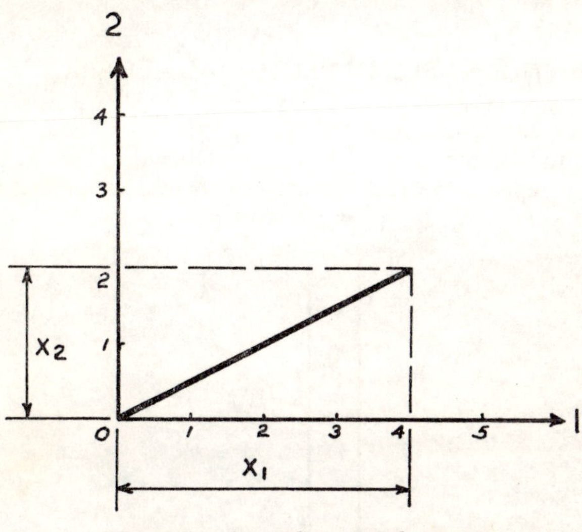

Fig. 2.1. A two dimensional vector.

In three dimensional space the components would be x_1, x_2 and x_3. More generally, in n dimensional space (hyper space) the vector would be specified as:

$$\mathbf{y} = (x_1, x_2, x_3, \ldots, x_n)$$

Hence any set of numbers or variables can be regarded as defining a vector. The number of components will depend upon the space dimensions but, for work in this book, we will normally deal with three dimensional vectors.

CYCLIC ORDER OF COMPONENTS

It is important that the components of vectors be stated in the same order. A set of three mutually orthogonal reference axes is shown in Figure 2.2. The convention generally in use is that the 01 (or $0X$) axis is drawn horizontal, the 02 (or $0Y$) axis is drawn vertical and the 03 (or $0Z$) axis is taken as normal to the plane of the paper. The components of vectors are listed in a cyclic order, starting with the component on the 01 axis and proceeding anticlockwise. If a three dimensional vector had components 3, 2 and 5 on the 01, 02 and 03 axes respectively, it would be written as (3, 2, 5).

It is important to remember that the order of the components must be consistent. A vector defined as (1, 2, 3) is an entirely different vector to the one defined as (3, 2, 1) when the same axes are used.

CONVENTIONS USED IN VECTORIAL ALGEBRA

It is often convenient to write $x_1, x_2, x_3, \ldots, x_n$ as x_i where i is assumed to have the values $1, 2, 3, \ldots, n$. Often the i itself is dropped. Present practice is to write the components of a vector as a column. Hence $\mathbf{y} = (x_1, x_2, x_3, \ldots, x_n)$ is written as:

$$\mathbf{y} = \begin{bmatrix} x_1 \\ x_2 \\ x_3 \\ \cdot \\ \cdot \\ \cdot \\ x_n \end{bmatrix} = [x_i] = [x]$$

A column of numbers representing a vector is enclosed in square brackets as illustrated. In some books and papers one still comes across the earlier convention which was to enclose vectors in curly brackets, $\{x_i\}$, but this practice is not now generally used.

SIGN OF COMPONENTS

For a set of mutually orthogonal reference axes the sign convention used in this book is the one in most general use and is illustrated in Figure 2.2.

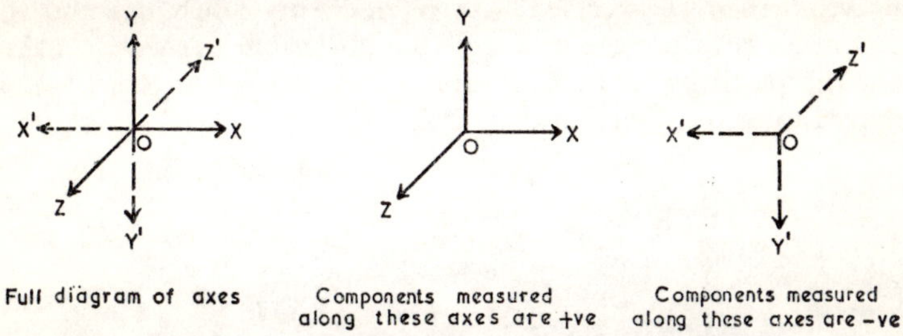

Fig. 2.2. Sign convention for components.

EQUALITY OF VECTORS

For two vectors to be equal not only their lengths but also their directions must be the same. This is only possible when all their components are equal.

LENGTH OF A VECTOR

For a vector in two dimensional space,

$$\mathbf{y} = \begin{bmatrix} x_1 \\ x_2 \end{bmatrix}$$

(See Figure 2.1)

The length, l, of the vector is given by $\sqrt{(x_1^2 + x_2^2)}$. For a vector in three dimensional space,

$$\mathbf{y} = \begin{bmatrix} x_1 \\ x_2 \\ x_3 \end{bmatrix}$$

$$l = \sqrt{(x_1^2 + x_2^2 + x_3^2)}$$

More generally, for a vector in n dimensional space, $\mathbf{y} = [x_i]$

$$l = \sqrt{\left(\sum_{i=1}^{i=n} x_i^2\right)}$$

DIRECTION OF A VECTOR

The length of a particular vector is a constant quantity but its direction is entirely dependent upon the set of reference axes used. Consider first a two dimensional vector (Figure 2.3). The direction of this vector can be specified with reference to either axis and the convention is to use the cosine of the angle that the vector makes with the particular axis. The value of the cosine of this angle is referred to as the direction cosine.

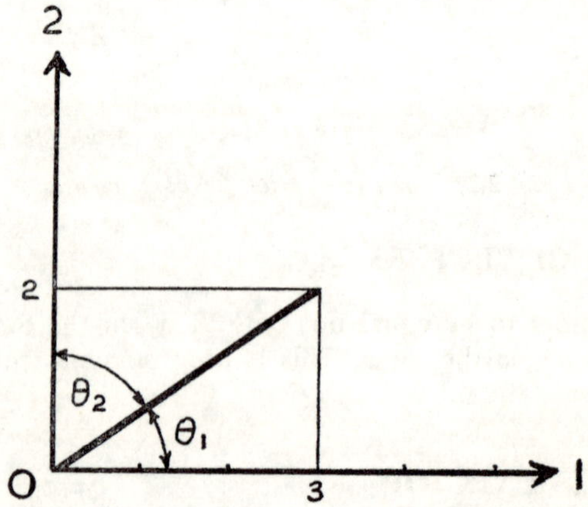

Fig. 2.3. *Direction of a two dimensional vector.*

In Figure 2.3 the reference axes are 01 and 02 and it is seen that the vector drawn is:

$$\mathbf{y} = \begin{bmatrix} 3 \\ 2 \end{bmatrix}$$

The direction cosine, referred to axis 01 is $\cos\theta_1 = x_1/l$
The direction cosine, referred to axis 02 is $\cos\theta_2 = x_2/l$

$$x_1 = 3; \quad x_2 = 2 \quad \text{and} \quad l = \sqrt{(3^2 + 2^2)} = \sqrt{13} = 3{\cdot}606$$

Hence $$\cos\theta_1 = \frac{x_1}{l} = \frac{3{\cdot}0}{3{\cdot}606} = 0{\cdot}832$$

And

$$\cos \theta_2 = \frac{x_2}{l} = \frac{2\cdot0}{3\cdot606} = 0\cdot555$$

From the foregoing it is seen that for a unit vector, i.e. a vector of unit length, the direction is completely specified by its components:

$$\cos \theta_1 = \frac{x_1}{l} = x_1 \quad \text{and} \quad \cos \theta_2 = \frac{x_2}{l} = x_2$$

Example

A vector, **v**, has components such that:

$$\mathbf{v} = \begin{bmatrix} 3 \\ 6 \\ 4 \end{bmatrix}$$

relative to three mutually orthogonal axes OX, OY and OZ. Determine the direction cosines of the vector.

There are three components of the vector hence the length will be:

$$l = \sqrt{(3^2 + 6^2 + 4^2)} = \sqrt{61} = 7\cdot81$$

$$\cos \theta_x = 3/7\cdot81 = 0\cdot384$$

$$\cos \theta_y = 6/7\cdot81 = 0\cdot768$$

$$\cos \theta_z = 4/7\cdot81 = 0\cdot512$$

MULTIPLICATION OF A VECTOR

If a vector is multiplied by a scalar quantity then each component of the vector is increased in proportion, i.e. the length of the vector will be different but not the direction.

$$k\mathbf{x} = k\begin{bmatrix} x_1 \\ x_2 \\ x_3 \\ \cdot \\ \cdot \\ \cdot \\ x_n \end{bmatrix} = \begin{bmatrix} kx_1 \\ kx_2 \\ kx_3 \\ \cdot \\ \cdot \\ \cdot \\ kx_n \end{bmatrix}$$

Example

A two dimensional vector has components 2 and 4 when referred to a particular set of orthogonal axes.

(i) Determine its length and direction.

(ii) If the vector is multiplied by the scalar quantity 4 determine its new length and direction.

(i) The vector is

$$\mathbf{y} = \begin{bmatrix} x_1 \\ x_2 \end{bmatrix} = \begin{bmatrix} 2 \\ 4 \end{bmatrix} \quad \text{Length} = \sqrt{(2^2 + 4^2)} = \sqrt{20} = 4{\cdot}472$$

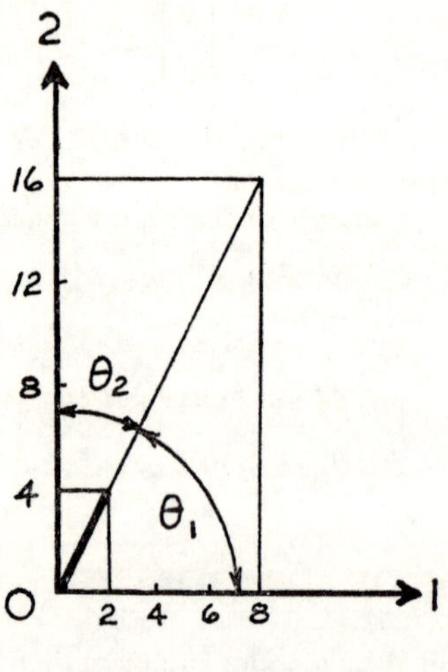

Fig. 2.4. *Example* 2.

The direction referred to axis 1 is $\cos\theta_1 = 2/4{\cdot}472 = 0{\cdot}447$, and referred to axis 2 is $\cos\theta_2 = 4/4{\cdot}472 = 0{\cdot}894$.

(ii) The vector is

$$4\mathbf{y} = 4\begin{bmatrix} 2 \\ 4 \end{bmatrix} = \begin{bmatrix} 8 \\ 16 \end{bmatrix} \quad \text{Length} = \sqrt{(8^2 + 16^2)} = 17{\cdot}9$$

The vector length is now 4 × 4·47 i.e. an increase of 4 times. The direction referred to axis 1 is $\cos\theta_1 = 8/17{\cdot}9 = 0{\cdot}447$, and referred to axis 2 is $\cos\theta_2 = 16/17{\cdot}9 = 0{\cdot}894$. The direction is unaltered.

Figure 2.4 illustrates the vector before and after multiplication.

NORMALISED VECTOR

A normalised vector is a unit vector, made up from another vector by dividing each component of the vector by its length.

Example

A vector is

$$\mathbf{x} = \begin{bmatrix} 2 \\ -4 \\ 3 \end{bmatrix}$$

determine its normalised vector.

$$\text{Length of vector} = \sqrt{(2^2 + 4^2 + 3^2)} = \sqrt{29}$$

$$\therefore \text{Normalised vector} = \frac{x}{\sqrt{29}} = \begin{bmatrix} 2/\sqrt{29} \\ -4/\sqrt{29} \\ 3/\sqrt{29} \end{bmatrix}$$

PRODUCT OF TWO VECTORS

If we have two vectors

$$\mathbf{a} = \begin{bmatrix} a_1 \\ a_2 \\ \cdot \\ \cdot \\ \cdot \\ a_n \end{bmatrix} \quad \text{and} \quad \mathbf{b} = \begin{bmatrix} b_1 \\ b_2 \\ \cdot \\ \cdot \\ \cdot \\ b_n \end{bmatrix}$$

then the product of vectors **a** and **b** is said to be the quantity **c** where $\mathbf{c} = a_1b_1 + a_2b_2 + a_3b_3 \cdots + a_nb_n$ (a single quantity).

Example

Determine the scalar product of the two vectors

$$\mathbf{y} = \begin{bmatrix} 1 \\ 6 \\ 8 \end{bmatrix} \quad \mathbf{x} = \begin{bmatrix} 2 \\ -3 \\ 5 \end{bmatrix}$$

$$c = 1 \times 2 - 6 \times 3 + 8 \times 5 = 24$$

ORTHOGONAL VECTORS

Vectors that are at right angles to each other are called orthogonal vectors. An important property of orthogonal vectors is that, when two are multiplied together their product is zero. This fact can be illustrated for the two dimensional vectors, $\mathbf{y}_1$ and $\mathbf{y}_2$ shown in Figure 2.5. The vectors have been plotted along the two reference axes 01 and 02 which are orthogonal.

Now

$$\mathbf{y}_1 = \begin{bmatrix} x_1 \\ 0 \end{bmatrix} \quad \mathbf{y}_2 = \begin{bmatrix} 0 \\ x_2 \end{bmatrix}$$

Hence

$$\mathbf{y}_1\mathbf{y}_2 = x_1 \times 0 + 0 \times x_2 = 0$$

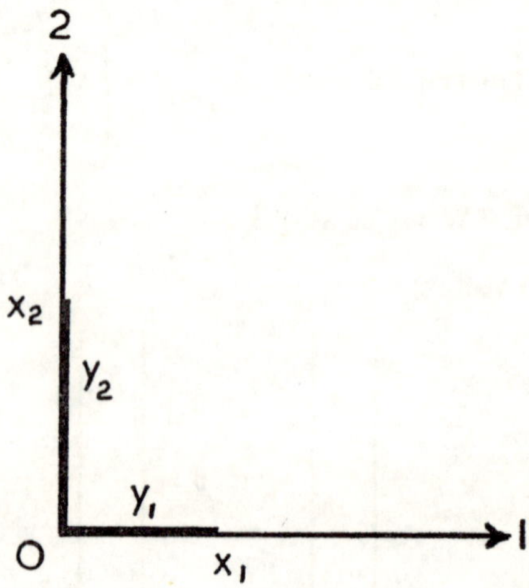

Fig. 2.5.

The argument can also be shown by the use of direction cosines. If we have two three dimensional vectors with direction cosines, l_1, m_1, n_1 and l_2, m_2, n_2 then the angle θ between them is given by:

$$\cos\theta = l_1l_2 + m_1m_2 + n_1n_2$$

In the special case of orthogonal vectors $\cos\theta = \cos 90° = 0$. Hence $l_1l_2 + m_1m_2 + n_1n_2 = 0$ which has been proved previously as a set of direction cosines can be considered as components of a unit vector.

Example

A vector is

$$\mathbf{v} = \begin{bmatrix} 2 \\ 1{\cdot}5 \end{bmatrix}$$

Determine the vectors that are of the same length and are orthogonal to it.

This is a two dimensional system. Therefore let the required vector, or vectors, be

$$\mathbf{y} = \begin{bmatrix} x_1 \\ x_2 \end{bmatrix}$$

For this vector (or vectors) to be orthogonal $\mathbf{yv} = 0$ i.e.

$$\begin{bmatrix} x_1 \\ x_2 \end{bmatrix} \begin{bmatrix} 2 \\ 1{\cdot}5 \end{bmatrix} = 0$$

$$\therefore \quad 2x_1 + 1{\cdot}5x_2 = 0$$

i.e.

$$x_1 = -0{\cdot}75x_2 \qquad (1)$$

The length of the new and the original vectors are the same:

$$\therefore \quad x_1{}^2 + x_2{}^2 = 2^2 + 1{\cdot}5^2$$

i.e.

$$x_1{}^2 + x_2{}^2 = 6{\cdot}25$$

Substitute for x_1 from (1)

$$(0{\cdot}75x_2)^2 + x_2{}^2 = 6{\cdot}25$$

$$\therefore \quad x_2 = \pm\, 2{\cdot}0$$

And, substituting in (1):

For

$$x_2 = +2{\cdot}0, \quad x_1 = -1{\cdot}5$$

For

$$x_2 = -2{\cdot}0, \quad x_1 = +1{\cdot}5$$

There are therefore two vectors that satisfy the required conditions

$$\mathbf{y}_1 = \begin{bmatrix} 1{\cdot}5 \\ -2{\cdot}0 \end{bmatrix} \quad \text{and} \quad \mathbf{y}_2 = \begin{bmatrix} -1{\cdot}5 \\ 2{\cdot}0 \end{bmatrix}$$

The three vectors are shown in Figure 2.6.

Example

If vector **v** in the previous example maintains its direction but is reduced to unit length determine its components.

Direction cosines of **v** are $\cos\theta_1$ and $\cos\theta_2$. Original length of

$$\mathbf{v} = \sqrt{6{\cdot}25} = 2{\cdot}5$$

$$\therefore \quad \cos\theta_1 = 2/2{\cdot}5 = 0{\cdot}8; \quad \cos\theta_2 = 1{\cdot}5/2{\cdot}5 = 0{\cdot}6$$

For a unit vector the direction cosines are equal to its components, hence

$$x_1 = 0{\cdot}8 \quad \text{and} \quad x_2 = 0{\cdot}6$$

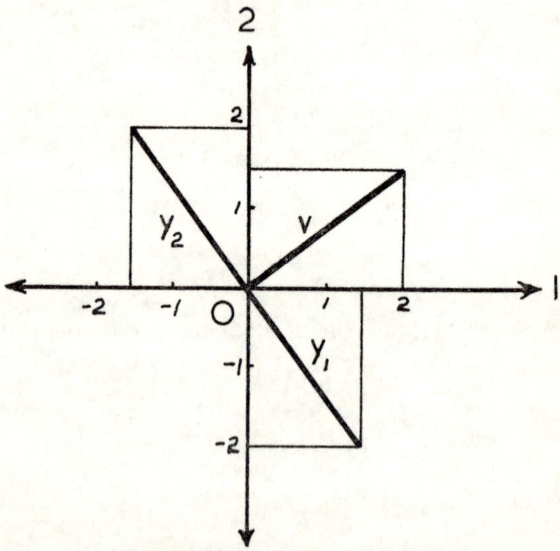

Fig. 2.6. *Example* 2.

Alternative Solution by simple proportion.
Original length = 2·5; Final length = 1·0
Hence

$$x_1 = 2 \times \frac{1{\cdot}0}{2{\cdot}5} = 0{\cdot}8 \qquad x_2 = 1{\cdot}5 \times \frac{1{\cdot}0}{2{\cdot}5} = 0{\cdot}6$$

Example

Two vectors are

$$\mathbf{y}_1 = \begin{bmatrix} 2{\cdot}0 \\ 1{\cdot}5 \\ 3{\cdot}0 \end{bmatrix} \quad \text{and} \quad \mathbf{y}_2 = \begin{bmatrix} 1{\cdot}5 \\ -2{\cdot}0 \\ 0{\cdot}0 \end{bmatrix}$$

Check that these two vectors are orthogonal and determine a third vector, of length 3, mutually orthogonal to them both.

If orthogonal then

$$\mathbf{y}_1\mathbf{y}_2 = 0$$

Now

$$\mathbf{y}_1\mathbf{y}_2 = 2{\cdot}0 \times 1{\cdot}5 - 1{\cdot}5 \times 2{\cdot}0 + 3{\cdot}0 \times 0{\cdot}0 = 0$$

The vectors are orthogonal. Let the unknown vector be:

$$\mathbf{y} = \begin{bmatrix} x_1 \\ x_2 \\ x_3 \end{bmatrix}$$

Then

$$\mathbf{y}_1\mathbf{y} = 0; \quad \mathbf{y}_2\mathbf{y} = 0 \quad \text{and} \quad l = \sqrt{(\Sigma x_i^2)}$$

$$\therefore \quad 2x_1 + 1{\cdot}5x_2 + 3x_3 = 0 \qquad (1)$$

$$1{\cdot}5x_1 - 2x_2 = 0 \qquad (2)$$

$$x_1^2 + x_2^2 + x_3^2 = 9 \qquad (3)$$

From (2) $x_1 = (4/3)x_2$ and, substituting in (3):

$$\frac{16}{9}x_2^2 + x_2^2 + x_3^2 = 9$$

i.e.

$$\frac{25}{9}x_2^2 + x_3^2 = 9 \qquad (4)$$

and substituting for x_1 in (1):

$$\frac{8}{3}x_2 + \frac{3}{2}x_2 + 3x_3 = 0$$

i.e.

$$x_2 = -\frac{18}{25}x_3$$

and, substituting in (4):

$$\frac{25}{9} \times \frac{18^2}{25^2}x_3^2 + x_3^2 = 9$$

i.e.

$$x_3 = \pm 1{\cdot}92$$

When

$$x_3 = +1{\cdot}92, \quad x_2 = -1{\cdot}38$$
$$x_3 = -1{\cdot}92, \quad x_2 = +1{\cdot}38$$

When

$$x_2 = -1{\cdot}38, \quad x_1 = -1{\cdot}845$$
$$x_2 = +1{\cdot}38, \quad x_1 = +1{\cdot}845$$

Hence the two vectors satisfying the conditions of the question are:

$$\mathbf{y}_3 = \begin{bmatrix} -1{\cdot}845 \\ -1{\cdot}38 \\ 1{\cdot}92 \end{bmatrix} \quad \text{and} \quad \mathbf{y}_4 = \begin{bmatrix} 1{\cdot}845 \\ 1{\cdot}38 \\ -1{\cdot}92 \end{bmatrix}$$

The positions of the four vectors are shown in Figure 2.7.

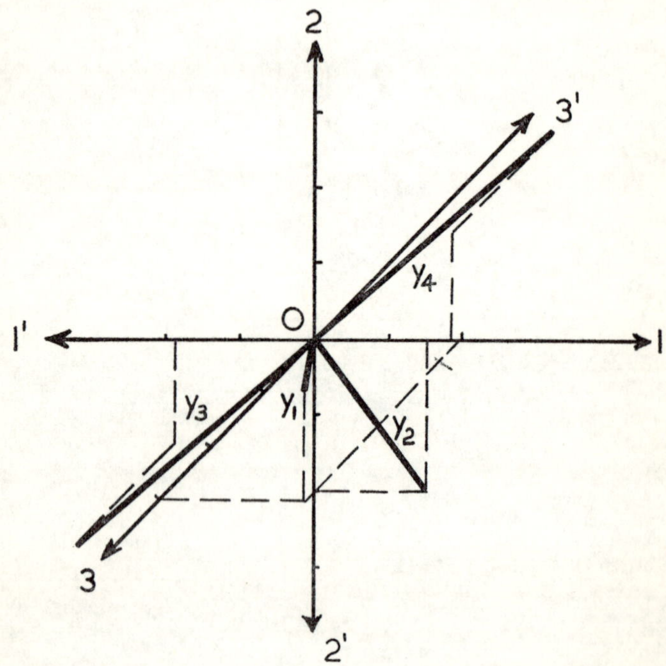

Fig. 2.7.

An example of the use of direction cosines arises from this problem. For $\mathbf{y}_1$, $l = \sqrt{(2^2 + 1{\cdot}5^2 + 3^2)} = 3{\cdot}9$. For $\mathbf{y}_2$, $l = 2{\cdot}5$. For $\mathbf{y}_3$ and $\mathbf{y}_4$, $l = 3$.

Vector	$\mathbf{y}_1$	$\mathbf{y}_2$	$\mathbf{y}_3$	$\mathbf{y}_4$
$\cos\theta_1$	$2/3{\cdot}9 = 0{\cdot}513$	$1{\cdot}5/2{\cdot}5 = 0{\cdot}6$	$-1{\cdot}845/3 = -0{\cdot}615$	$0{\cdot}615$
$\cos\theta_2$	$1{\cdot}5/3{\cdot}9 = 0{\cdot}385$	$-2/2{\cdot}5 = -0{\cdot}8$	$-1{\cdot}38/3 = -0{\cdot}46$	$0{\cdot}46$
$\cos\theta_3$	$3/3{\cdot}9 = 0{\cdot}769$	0	$1{\cdot}92/3 = 0{\cdot}64$	$-0{\cdot}64$

These direction cosines, $\cos\theta_1$, $\cos\theta_2$, $\cos\theta_3$ can be considered as x_1, x_2 and x_3 components of a set of unit vectors lying in the same directions as vectors $\mathbf{y}_1$, $\mathbf{y}_2$, $\mathbf{y}_3$ and $\mathbf{y}_4$.

The reader might like to check that the product of any two sets of these components is zero, proving that the vectors $\mathbf{y}_1$, $\mathbf{y}_2$, $\mathbf{y}_3$ and $\mathbf{y}_4$ are mutually orthogonal.

Note The symbol for $\cos\theta$ is generally abbreviated to l, m and n, taken in cyclic order.
i.e. in a unit vector:

$$\cos\theta_1 = l; \quad \cos\theta_2 = m; \quad \cos\theta_3 = n$$

A PLANE

A plane has dimensions in two directions only and may be defined as a surface on which the straight line joining any two points on it lies on the surface.

The general equation of a plane is given by the first degree expression:

$$ax + by + cz + d = 0$$

Let points $A(x_1, y_1, z_1)$ and $B(x_2, y_2, z_2)$ be two points on a surface and let $P(x, y, z)$ be a third point lying on the straight line AB such that AP $= s$ and PB $= t$.
i.e.

$$\frac{\text{AP}}{\text{PB}} = \frac{s}{t}$$

If $ax + by + cz + d = 0$ is the general equation of a plane then it must satisfy both points A and B. i.e.:

$$ax_1 + by_1 + cz_1 + d = 0 \tag{1}$$

$$ax_2 + by_2 + cz_2 + d = 0 \tag{2}$$

Multiplying equation (1) by $t/(s + t)$ and equation (2) by $s/(s + t)$ and adding:

$$a\left(\frac{tx_1 + sx_2}{s + t}\right) + b\left(\frac{ty_1 + sy_2}{s + t}\right) + c\left(\frac{tz_1 + sz_2}{s + t}\right) + d\left(\frac{t + s}{s + t}\right) = 0$$

Now, by proportion:

$$\frac{x - x_1}{x_2 - x} = \frac{s}{t}$$

therefore
$$tx - tx_1 = sx_2 - sx$$
$$x(s + t) = tx_1 + sx_2$$
$$x = \frac{tx_1 + sx_2}{s + t}$$

Similar expressions can be derived for y and z. Hence
$$ax + by + cz + d = 0$$

P lies on the surface no matter what the value of s and t and therefore satisfies the requirements of a plane. The equation $ax + by + cz + d = 0$ is the equation of a plane.

PLANE PASSING THROUGH THE ORIGIN

When a plane passes through the origin then the coordinates of the origin (0, 0, 0) must satisfy the equation $ax + by + cz + d = 0$ i.e.
$$a0 + b0 + c0 + d = 0$$
Hence
$$d = 0$$
the equation of a plane passing through the origin is:
$$ax + by + cz = 0$$

LINEAR DEPENDENCE

As has been shown, if two vectors
$$\begin{bmatrix} x_1 \\ y_1 \\ z_1 \end{bmatrix} \quad \text{and} \quad \begin{bmatrix} x_3 \\ y_2 \\ z_2 \end{bmatrix}$$
are orthogonal then their product $c = x_1x_2 + y_1y_2 + z_1z_2 = 0$.

This is the equation of a plane passing through the origin, in other words a two dimensional surface lying in a three dimensional space defined by the reference axes OX, OY and OZ.

Any vector whose components satisfy this equation must lie in the plane. With this condition to be satisfied such a vector need have only two of its components known for it to be completely defined. The three components of the vector are linearly dependent.

The argument applies in an n-dimensional space system.

If the components of such vectors have to satisfy an equation of the form:

$$a_1x_1 + a_2x_2 + \ldots + a_nx_n = 0$$

then they are linearly dependent in that only $(n - 1)$ components can be given arbitrary values. The vector is restricted to a space of $(n - 1)$ dimensions lying in an n-dimensional system. This restricted space is often referred to as a hyperplane.

3

Matrices

A matrix is simply a collection of numbers, which may be real or complex. For example:

$$\begin{bmatrix} a & b & c \\ d & e & f \end{bmatrix} \quad \text{or} \quad \begin{bmatrix} 4 & 5 & 6 \\ 7 & 8 & 9 \\ 1 & 2 & 3 \end{bmatrix}$$

Unlike determinants, matrices need not have the same number of rows as columns and can be either square or rectangular. The number of rows is generally given the symbol m and the number of columns the symbol n.

Example

$$\begin{bmatrix} 6 & -5 & 12 \\ -8 & 2 & 3 \end{bmatrix}$$

There are 2 rows $\quad \therefore m = 2, \qquad$ and 3 columns $\quad \therefore n = 3.$

This matrix is referred to as a 2×3 matrix. A matrix with m rows and n columns is a $m \times n$ matrix.

ELEMENT OF A MATRIX

Each term in a matrix is called an element. The convention used for a general element is the symbol a_{ij}, i.e. the element in row i and column j. As can be seen from the above examples of matrices a_{ij} is not necessarily equal to a_{ji}.

Generally a matrix is used in its entirety and, to save writing the matrix out in full the convention is to refer to it by a single upper case bold letter, **A**, **B** or **C**, etc. Vectors, it will be remembered, are referred to by single lower case bold letters, **a**, **b** or **c**, etc.

SQUARE MATRIX

A square matrix is one with the number of rows equal to the number of columns, i.e. $m = n$.

Example

$$\mathbf{A} = \begin{bmatrix} 1 & 2 & 3 \\ 4 & 5 & 6 \\ 7 & 8 & 9 \end{bmatrix}$$

$$\mathbf{B} = \begin{bmatrix} a_{11} & a_{12} & a_{13} \\ a_{21} & a_{22} & a_{23} \\ a_{31} & a_{32} & a_{33} \end{bmatrix}$$

Both the above matrices are 3 by 3 (square) matrices.

It should be noted that whereas determinants are grouped between two straight lines a matrix is grouped between a pair of square brackets. These brackets are used when the matrix is written out in full but are often dropped when the symbol **A**, **B**, etc. is used.

LEADING ELEMENTS

The elements a_{ii} for all values of i.

Example

$$\begin{bmatrix} a_{11} & a_{12} & a_{13} & a_{14} \\ a_{21} & a_{22} & a_{23} & a_{24} \\ a_{31} & a_{32} & a_{33} & a_{34} \\ a_{41} & a_{42} & a_{43} & a_{44} \end{bmatrix}$$

The elements a_{11}, a_{22}, a_{33} and a_{44} are leading elements.

LEADING DIAGONAL

This occurs in a square matrix and is the line consisting of the leading elements, i.e. the line from the top left hand corner of the matrix to the bottom right hand corner.

TRACE OF A MATRIX

The sum of the elements of the leading diagonal.

DIAGONAL MATRIX

A matrix in which all the elements are zero except those on the leading diagonal, e.g.

$$\begin{bmatrix} 1 & 0 & 0 & 0 \\ 0 & 2 & 0 & 0 \\ 0 & 0 & 3 & 0 \\ 0 & 0 & 0 & 4 \end{bmatrix}$$

UNIT MATRIX

A diagonal matrix in which the nonzero elements are all equal to unity, e.g.

$$\begin{bmatrix} 1 & 0 & 0 & 0 \\ 0 & 1 & 0 & 0 \\ 0 & 0 & 1 & 0 \\ 0 & 0 & 0 & 1 \end{bmatrix}$$

Usually a unit matrix is given the symbol **I**.

NULL MATRIX

A matrix in which all the elements are equal to zero. Generally this matrix is given the symbol **O**.

MATRICES WITH A LARGE NUMBER OF ZERO ELEMENTS

In engineering it often happens that a square matrix may have a large number of nonzero elements. Such matrices can often be classified as one or other of the three types listed below.

BANDED OR BAND MATRIX

In this square matrix all elements are zero except those along a strip often, but not always, centred along the leading diagonal.

Example

$$\mathbf{A} = \begin{bmatrix} a_{11} & a_{12} & 0 & 0 & 0 & 0 \\ a_{21} & a_{22} & a_{23} & 0 & 0 & 0 \\ 0 & a_{32} & a_{33} & a_{34} & 0 & 0 \\ 0 & 0 & a_{43} & a_{44} & a_{45} & 0 \\ 0 & 0 & 0 & a_{54} & a_{55} & a_{56} \\ 0 & 0 & 0 & 0 & a_{65} & a_{66} \end{bmatrix}$$

In this example the band is centred on the leading diagonal and has a width of three elements. Such a matrix would be called a tridiagonal banded matrix.

This form of matrix is encountered in the finite element method of analysis.

UPPER TRIANGULAR MATRIX

In this square matrix all elements below the leading diagonal are zero.

Example

$$\mathbf{A} = \begin{bmatrix} a_{11} & a_{12} & a_{13} & a_{14} & a_{15} \\ 0 & a_{22} & a_{23} & a_{24} & a_{25} \\ 0 & 0 & a_{33} & a_{34} & a_{35} \\ 0 & 0 & 0 & a_{44} & a_{45} \\ 0 & 0 & 0 & 0 & a_{55} \end{bmatrix}$$

LOWER TRIANGULAR MATRIX

In this square matrix all elements above the leading diagonal are zero.

Example

$$\mathbf{A} = \begin{bmatrix} a_{11} & 0 & 0 & 0 & 0 \\ a_{21} & a_{22} & 0 & 0 & 0 \\ a_{31} & a_{32} & a_{33} & 0 & 0 \\ a_{41} & a_{42} & a_{43} & a_{44} & 0 \\ a_{51} & a_{52} & a_{53} & a_{54} & a_{55} \end{bmatrix}$$

MULTIPLICATION OF MATRICES

As will be seen later there are many instances in stress analysis where one set of variables is related to another set, i.e. when one vector is related to another, by a set of linear equations (equations of the first degree).

We may have two vectors:

$$\mathbf{x} = \begin{bmatrix} x_1 \\ x_2 \\ x_3 \\ . \\ . \\ . \\ x_m \end{bmatrix} \quad \text{and} \quad \mathbf{y} = \begin{bmatrix} y_1 \\ y_2 \\ y_3 \\ . \\ . \\ . \\ y_n \end{bmatrix}$$

which are related to each other by equations of the form:

$$x_1 = a_{11}y_1 + a_{12}y_2 + a_{13}y_3 + \ldots + a_{1n}y_n \qquad (1)$$

$$x_2 = a_{21}y_1 + a_{22}y_2 + a_{23}y_3 + \ldots + a_{2n}y_n \qquad (2)$$

$$\begin{matrix} . & . & . & . & . \\ . & . & . & . & . \\ . & . & . & . & . \end{matrix}$$

$$x_m = a_{m1}y_1 + a_{m2}y_2 + a_{m3}y_3 + \ldots + a_{mn}y_n$$

the coefficients a_{11}, a_{12}, etc. being constants.

These relationships can be written in matrix form:

$$\begin{bmatrix} x_1 \\ x_2 \\ . \\ . \\ . \\ x_m \end{bmatrix} = \begin{bmatrix} a_{11} & a_{12} & a_{13} & \ldots & a_{1n} \\ a_{21} & a_{22} & a_{23} & \ldots & a_{2n} \\ . & . & . & \ldots & . \\ . & . & . & \ldots & . \\ . & . & . & \ldots & . \\ a_{m1} & a_{m2} & a_{m3} & \ldots & a_{mn} \end{bmatrix} \begin{bmatrix} y_1 \\ y_2 \\ . \\ . \\ . \\ y_n \end{bmatrix}$$

or, using the general symbols:

$$[\mathbf{x}] = [\mathbf{A}][\mathbf{y}]$$

or, more usually:

$$\mathbf{x} = \mathbf{A}\mathbf{y}$$

In other words the relationship between the two vectors can be said to be that vector **x** is equal to vector **y** multiplied by matrix **A**.

To yield equation (1) from the matrix form of the relationship it is seen that the first row of the matrix **A** must be placed alongside the vector **y** and each corresponding term multiplied together:

$$x_1 = [a_{11} \quad a_{12} \quad a_{13} \quad \cdots \quad a_{1n}] \begin{bmatrix} y_1 \\ y_2 \\ y_3 \\ \cdot \\ \cdot \\ \cdot \\ y_n \end{bmatrix}$$

A good way to remember this technique is to think of a swimmer diving off a board. A horizontal move forward and then the vertical dive:

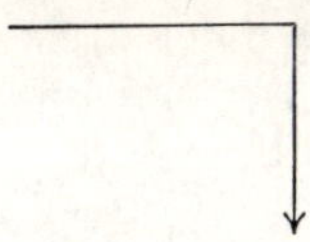

Remember it is the row in the matrix that does the 'dive'.

Once this move has been accomplished the two sets of figures will lie alongside each other:

$$x_1 = \begin{bmatrix} y_1 \\ y_2 \\ y_3 \\ \cdot \\ \cdot \\ \cdot \\ y_n \end{bmatrix} \begin{bmatrix} a_{11} \\ a_{12} \\ a_{13} \\ \cdot \\ \cdot \\ \cdot \\ a_{1n} \end{bmatrix}$$

When the corresponding terms are multiplied together and the resulting elements added we have equation (1):

$$x_1 = a_{11}y_1 + a_{12}y_2 + a_{13}y_3 + \ldots + a_{1n}y_n$$

Hence we see that in order to obtain the *first* term of vector **x** multiply vector **y** by the *first* row of matrix **A**.

The reader may like to check that to obtain the second term of vector **x**, i.e. x_2 one must multiply vector **y** by the *second* row of matrix **A** (this procedure will of course yield equation (2)).

For the multiplication to take place the number of terms in both sets of figures must be equal, in other words in order that two matrices can be multiplied together the number of columns in the first matrix must equal the number of rows in the second. Similarly when a vector is multiplied by a matrix the number of columns in the matrix must be equal to the number of elements in the vector.

Example

$$\begin{bmatrix} 1 & 2 & 3 \\ 4 & 5 & 6 \end{bmatrix} \begin{bmatrix} 1 \\ 2 \\ 3 \end{bmatrix} = \mathbf{y} \qquad \text{Determine } \mathbf{y}.$$

This is an example of a vector multiplied by a matrix. The product must have the same number of columns as the original vector and the same number of rows as the matrix. Hence the product will be a vector of two terms:
Let vector **y** be represented by

$$\begin{bmatrix} y_1 \\ y_2 \end{bmatrix}$$

Then to find y_1, the *first* term, multiply the original vector by the *first* row of the matrix. Remember the 'diving' procedure:

$$y_1 = \begin{bmatrix} 1 \\ 2 \\ 3 \end{bmatrix} \begin{bmatrix} 1 \\ 2 \\ 3 \end{bmatrix} = 1 \times 1 + 2 \times 2 + 3 \times 3 = 14$$

To find y_2, the *second* term of the vector multiply the original vector by the *second* row of the matrix:

$$y_2 = 1 \times 4 + 2 \times 5 + 3 \times 6 = 32$$

Hence

$$\mathbf{y} = \begin{bmatrix} 14 \\ 32 \end{bmatrix}$$

As the reader will appreciate most of the above work can be done without writing out the whole procedure formally.

Example

$$\begin{bmatrix} 4 & 3 & 2 \\ 1 & 1 & 1 \\ 0 & 2 & 8 \end{bmatrix} \begin{bmatrix} 6 & 3 & 1 \\ 1 & 5 & 2 \\ 2 & 9 & 4 \end{bmatrix} = \mathbf{A} \qquad \text{Determine } \mathbf{A}.$$

This is an example of two matrices multiplied together to yield a matrix **A**. The expression is of the form **BC** = **A**. Matrix **B** has 3 columns and matrix **C** has 3 rows, therefore multiplication is possible. *First* term of first column of matrix **A** = a_{11}. This term is found by multiplying the *first* column of **C** by the *first* row of **B**. Hence the element a_{11} in matrix **A** is equal to $4 \times 6 + 3 \times 1 + 2 \times 2 = 31$. Similarly *second* term of *first* column of matrix **A**(a_{21}) = *first* column of matrix **C** by the *second* row of matrix **B**.
i.e.

$$a_{21} = 1 \times 6 + 1 \times 1 + 1 \times 2 = 9$$

Similarly

$$a_{31} = 0 \times 6 + 2 \times 1 + 8 \times 2 = 18$$

To obtain the *first* term of the *second* column of matrix **A**(a_{12}) we must multiply the *second* column of matrix **C** by the *first* row of matrix **B**, i.e.

$$a_{12} = 4 \times 3 + 3 \times 5 + 2 \times 9 = 45$$

Similarly

$$a_{22} = 1 \times 3 + 1 \times 5 + 1 \times 9 = 17$$

and

$$a_{32} = 82$$

A similar procedure will yield the third column of matrix **A** and the resulting product is:

$$\begin{bmatrix} 4 & 3 & 2 \\ 1 & 1 & 1 \\ 0 & 2 & 8 \end{bmatrix} \begin{bmatrix} 6 & 3 & 1 \\ 1 & 5 & 2 \\ 2 & 9 & 4 \end{bmatrix} = \begin{bmatrix} 31 & 45 & 18 \\ 9 & 17 & 7 \\ 18 & 82 & 36 \end{bmatrix}$$

Example

$$\begin{bmatrix} 1 & 0 & 0 \\ 0 & 1 & 0 \\ 0 & 0 & 1 \end{bmatrix} \begin{bmatrix} 2 & 3 \\ 4 & 5 \\ 6 & 7 \end{bmatrix} = \mathbf{A} \qquad \text{Determine } \mathbf{A}.$$

If the reader applies the rules of multiplication that have just been developed he will find that matrix **A** is equal to the original matrix

$$\begin{bmatrix} 2 & 3 \\ 4 & 5 \\ 6 & 7 \end{bmatrix}$$

before it was multiplied by the unit matrix

$$\begin{bmatrix} 1 & 0 & 0 \\ 0 & 1 & 0 \\ 0 & 0 & 1 \end{bmatrix}$$

This is an important law that we have established. Multiplication of a vector or a matrix by a unit matrix leaves the original vector or matrix unaltered.

FURTHER NOTES ON MATRIX MULTIPLICATION

It may have been noted that in all the matrix multiplications so far the operations were carried out working from left to right. This is essential. In a multiplication such as **C** = **AB** the matrix **A** is said to premultiply the matrix **B** whereas matrix **B** is said to post multiply matrix **A**. Although **C** = **AB** it does not necessarily follow that **C** = **BA**.

Example

Determine **D** = **ABC** where

$$\mathbf{A} = \begin{bmatrix} 1 & 2 \\ 3 & 4 \end{bmatrix} \quad \mathbf{B} = \begin{bmatrix} 3 & 4 \\ 5 & 6 \end{bmatrix} \quad \mathbf{C} = \begin{bmatrix} 5 & 6 \\ 7 & 8 \end{bmatrix}$$

Remember that the working must be carried out from left to right.

Hence

$$\mathbf{D} = \mathbf{A}(\mathbf{BC}) \quad \text{or} \quad \mathbf{D} = (\mathbf{AB})\mathbf{C}$$

$$\mathbf{D} = \mathbf{A}(\mathbf{BC}) = \begin{bmatrix} 1 & 2 \\ 3 & 4 \end{bmatrix} \begin{bmatrix} 43 & 50 \\ 67 & 78 \end{bmatrix} = \begin{bmatrix} 177 & 206 \\ 397 & 462 \end{bmatrix}$$

$$\mathbf{D} = (\mathbf{AB})\mathbf{C} = \begin{bmatrix} 13 & 16 \\ 29 & 36 \end{bmatrix} \begin{bmatrix} 5 & 6 \\ 7 & 8 \end{bmatrix} = \begin{bmatrix} 177 & 206 \\ 397 & 462 \end{bmatrix}$$

It is interesting to note that $\mathbf{D} \neq (\mathbf{CA})\mathbf{B}$

$$(\mathbf{CA})\mathbf{B} = \begin{bmatrix} 23 & 34 \\ 31 & 46 \end{bmatrix} \begin{bmatrix} 3 & 4 \\ 5 & 6 \end{bmatrix} = \begin{bmatrix} 239 & 296 \\ 323 & 400 \end{bmatrix}$$

This numerical example illustrates an important law in matrix algebra, i.e. matrices are associative but noncommutative, i.e. **AB** is not necessarily equal to **BA**. This departure from the usual algebraic convention is of little consequence provided that when carrying out multiplication the operations are always carried out from left to right.

ADDITION AND SUBTRACTION OF MATRICES

This is a relatively simple operation in that the corresponding elements of the two matrices are added (or subtracted) to yield the element in the resultant matrix.

Example

$$\begin{bmatrix} 1 & 2 & 3 \\ 4 & 5 & 6 \end{bmatrix} + \begin{bmatrix} 2 & 4 & 8 \\ 5 & 3 & 2 \end{bmatrix} = \begin{bmatrix} 3 & 6 & 11 \\ 9 & 8 & 8 \end{bmatrix}$$

$$\begin{bmatrix} 1 & 2 & 3 \\ 4 & 5 & 6 \end{bmatrix} - \begin{bmatrix} 2 & 4 & 8 \\ 5 & 3 & 2 \end{bmatrix} = \begin{bmatrix} -1 & -2 & -5 \\ -1 & 2 & 4 \end{bmatrix}$$

TRANSPOSE

Associated with any matrix **A** is another matrix $\mathbf{A}^t$ known as its transpose. The transpose matrix is formed by changing the rows into columns and the columns into rows.

Example

$$\mathbf{A} = \begin{bmatrix} 5 & 6 & 7 & 8 \\ 1 & 2 & 3 & 4 \\ 5 & 6 & 7 & 8 \end{bmatrix} \quad \mathbf{A}^t = \begin{bmatrix} 5 & 1 & 5 \\ 6 & 2 & 6 \\ 7 & 3 & 7 \\ 8 & 4 & 8 \end{bmatrix}$$

By selecting suitable matrices the reader can verify a useful law that is associated with transposes:

$$(\mathbf{AB})^t = \mathbf{B}^t\mathbf{A}^t$$

In a more general form the law becomes:

$$(\mathbf{A}_1\mathbf{A}_2\mathbf{A}_3 \ldots \mathbf{A}_n)^t = \mathbf{A}_n{}^t\mathbf{A}_{n-1}{}^t \ldots \mathbf{A}_3{}^t\mathbf{A}_2{}^t\mathbf{A}_1{}^t$$

SYMMETRIC MATRIX

A matrix that is equal to its transpose is called a symmetric matrix.

Example

$$\mathbf{A} = \begin{bmatrix} 2 & 3 & 4 \\ 3 & 5 & 6 \\ 4 & 6 & 7 \end{bmatrix} \quad \mathbf{A}^t = \begin{bmatrix} 2 & 3 & 4 \\ 3 & 5 & 6 \\ 4 & 6 & 7 \end{bmatrix}$$

A symmetric matrix must be square.

It should be noted that the product of two symmetric matrices is not itself necessarily symmetric.

MULTIPLICATION OF VECTORS

Two vectors, **x** and **y**, when written in their normal column vector form cannot be multiplied by the rules that we have established, e.g.

$$\begin{bmatrix} x_1 \\ x_2 \\ x_3 \\ \cdot \\ \cdot \\ \cdot \\ x_n \end{bmatrix} \begin{bmatrix} y_1 \\ y_2 \\ y_3 \\ \cdot \\ \cdot \\ \cdot \\ y_n \end{bmatrix}$$

cannot be multiplied together. It is necessary to write the vectors in the form:

$$[x_1 x_2 x_3 \ldots x_n] \begin{bmatrix} y_1 \\ y_2 \\ y_3 \\ \cdot \\ \cdot \\ \cdot \\ y_n \end{bmatrix}$$

Multiplication is now possible (provided that the number of columns in the first vector equals the number of rows in the second).

Hence in order to obtain the product **xy** we must use $\mathbf{x}^t\mathbf{y}$. Using the transpose of **x** does not alter the vector in any way, the technique is carried out simply to make multiplication possible.

It is obvious that $\mathbf{x}^t\mathbf{y} = \mathbf{y}^t\mathbf{x}$. This is not the case with matrices and the rule can only be applied to column vectors.

ROTATION OF A VECTOR

Consider a unit vector with components x_1 and x_2 (Figure 3.1A)

Let the vector rotate through an angle α such that θ_1 becomes θ_1' and θ_2 becomes θ_2' (Figure 3.1B)

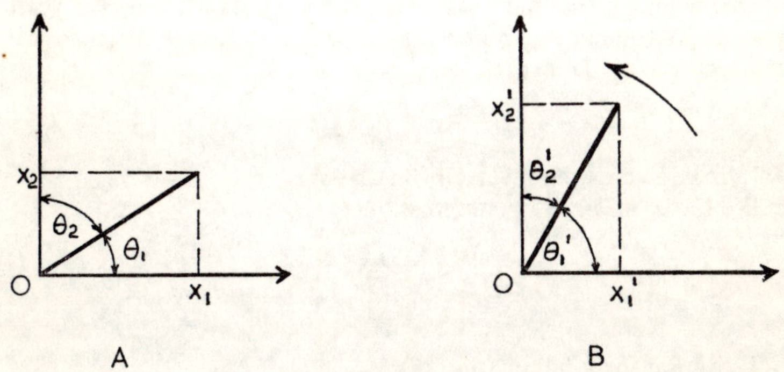

Fig. 3.1.

Then

$$\theta_1' = \theta_1 + \alpha$$

$$\theta_2' = \theta_2 - \alpha$$

As the vector has unit length $\cos\theta_1 = x_1$, $\cos\theta_2 = \cos(90° - \theta_1) = \sin\theta_1 = x_2$

$$x_1' = \cos\theta_1' = \cos(\theta_1 + \alpha) = \cos\theta_1 \cos\alpha - \sin\theta_1 \sin\alpha$$

$$= x_1 \cos\alpha - x_2 \sin\alpha$$

$$x_2' = \cos\theta_2' = \cos(\theta_2 - \alpha) = \cos\theta_2 \cos\alpha + \sin\theta_2 \sin\alpha$$

$$= x_2 \cos\alpha + x_1 \sin\alpha$$

$$= x_1 \sin\alpha + x_2 \cos\alpha$$

i.e.

$$\begin{bmatrix} \cos\alpha & -\sin\alpha \\ \sin\alpha & \cos\alpha \end{bmatrix} \begin{bmatrix} x_1 \\ x_2 \end{bmatrix} = \begin{bmatrix} x_1' \\ x_2' \end{bmatrix}$$

i.e. the matrix

$$\begin{bmatrix} \cos\alpha & -\sin\alpha \\ \sin\alpha & \cos\alpha \end{bmatrix}$$

produces a rotation of angle α.

INVERSE MATRIX

The expression $\mathbf{y} = \mathbf{A}\mathbf{x}$ represents a linear transformation of an arbitrary vector $\mathbf{x}$ into another vector $\mathbf{y}$. The operation is carried out by multiplying vector $\mathbf{x}$ by the matrix $\mathbf{A}$.

One would expect that a reverse process exists whereby the vector $\mathbf{y}$, when multiplied by a matrix that is some function of $\mathbf{A}$ would yield the vector $\mathbf{x}$ by the expression

$$\mathbf{x} = \mathbf{A}^{-1}\mathbf{y}$$

$\mathbf{A}^{-1}$ is known as the inverse of matrix $\mathbf{A}$.

We therefore have two expressions:

$$\mathbf{y} = \mathbf{A}\mathbf{x} \qquad (1)$$

$$\mathbf{x} = \mathbf{A}^{-1}\mathbf{y} \qquad (2)$$

Substituting for $\mathbf{y}$ in equation (2);

$$\mathbf{x} = \mathbf{A}\mathbf{A}^{-1}\mathbf{x}$$

This means that the product $\mathbf{AA}^{-1}$ must yield a matrix which does not alter $\mathbf{x}$ when the two are multiplied together.

The only matrix with this property is the unit matrix, which can either pre- or post-multiply another matrix or vector without altering any values.

Hence

$$\mathbf{AA}^{-1} = \mathbf{A}^{-1}\mathbf{A} = \mathbf{I}$$

A matrix when multiplied by its inverse will yield a unit matrix. This can often be a useful check during the evaluation of an inverse. A rule concerning the inverse of the product of two square matrices $\mathbf{A}$ and $\mathbf{B}$, is that $(\mathbf{AB})^{-1} = \mathbf{B}^{-1}\mathbf{A}^{-1}$. This can be shown to be so by multiplying $\mathbf{AB}$ by $\mathbf{B}^{-1}\mathbf{A}^{-1}$

$$\mathbf{B}^{-1}\mathbf{A}^{-1}\mathbf{AB} = \mathbf{B}^{-1}\mathbf{IB} = \mathbf{B}^{-1}\mathbf{B} = \mathbf{I}$$

Hence

$$(\mathbf{AB})^{-1} = \mathbf{B}^{-1}\mathbf{A}^{-1}$$

In a general form the law becomes:

$$(\mathbf{A}_1\mathbf{A}_2\mathbf{A}_3 \ldots \mathbf{A}_n)^{-1} = \mathbf{A}_n{}^{-1}\mathbf{A}^{-1}{}_{n-1} \ldots \mathbf{A}_3{}^{-1}\mathbf{A}_2{}^{-1}\mathbf{A}_1{}^{-1}$$

where all the matrices are square.

EVALUATION OF THE INVERSE MATRIX

Essentially in finding the inverse of a matrix one must solve the equation $\mathbf{AA}^{-1} = \mathbf{I}$, $\mathbf{A}$ and $\mathbf{I}$ being known, i.e. one is solving a set of simultaneous equations.

In order to examine the procedure one must look at two other forms of a matrix.

COFACTOR MATRIX

This matrix is formed by replacing every element of the original matrix by its corresponding cofactor. The symbol generally used for the cofactor matrix is $\mathbf{A}^c$.

Example

$$\mathbf{A} = \begin{bmatrix} 4 & 3 & 2 \\ 1 & 8 & 4 \\ 6 & 3 & 5 \end{bmatrix} \quad \text{Determine } \mathbf{A}^c.$$

Consider a_{11}.

$$\text{Cofactor} = +\begin{vmatrix} 8 & 4 \\ 3 & 5 \end{vmatrix} = 28$$

(The rules for the sign of a cofactor were established in chapter 1.) Similarly for the other elements. Hence:

$$\mathbf{A}^{c} = \begin{bmatrix} 28 & 19 & -45 \\ -9 & 8 & 6 \\ -4 & -14 & 29 \end{bmatrix}$$

ADJOINT MATRIX, $\mathbf{A}^{a}$

The adjoint matrix is simply the transpose of the cofactor matrix.

Example

Considering the previous matrix:

$$\mathbf{A}^{a} = (\mathbf{A}^{c})^{t} = \begin{bmatrix} 28 & -9 & -4 \\ 19 & 8 & -14 \\ -45 & 6 & 29 \end{bmatrix}$$

The adjoint matrix is closely related to the inverse matrix as follows.

Consider the product

$$\mathbf{AA}^{a} = \begin{bmatrix} a_{11} & a_{12} & a_{13} \\ a_{21} & a_{22} & a_{23} \\ a_{31} & a_{32} & a_{33} \end{bmatrix} \begin{bmatrix} a_{11}^{c} & a_{21}^{c} & a_{31}^{c} \\ a_{12}^{c} & a_{22}^{c} & a_{32}^{c} \\ a_{13}^{c} & a_{23}^{c} & a_{33}^{c} \end{bmatrix}$$

Let

$$\mathbf{AA}^{a} = \mathbf{B} = \begin{bmatrix} b_{11} & b_{12} & b_{13} \\ b_{21} & b_{22} & b_{23} \\ b_{31} & b_{32} & b_{33} \end{bmatrix}$$

Now

$$b_{11} = [a_{11}a_{12}a_{13}] \begin{bmatrix} a_{11}^{c} \\ a_{12}^{c} \\ a_{13}^{c} \end{bmatrix}$$

The term b_{11} therefore consists of the elements of the first row of the original matrix multiplied by their corresponding cofactors, i.e. b_{11} is simply equal to the determinant of the original matrix $|\mathbf{A}|$.

$$b_{12} = [a_{11}a_{12}a_{13}] \begin{bmatrix} a_{21}^{\mathrm{c}} \\ a_{22}^{\mathrm{c}} \\ a_{23}^{\mathrm{c}} \end{bmatrix}$$

i.e. b_{12} is made up by taking the elements of the first row of the original matrix and multiplying them by the cofactors of the second row and adding. As was established on page 10 such an expansion is equal to zero.

If all the elements are considered in turn it is seen that:

$$\mathbf{AA}^{\mathrm{a}} = \begin{bmatrix} |\mathbf{A}| & 0 & 0 \\ 0 & |\mathbf{A}| & 0 \\ 0 & 0 & |\mathbf{A}| \end{bmatrix}$$

and, introducing the unit matrix:

$$\mathbf{I} = \begin{bmatrix} 1 & 0 & 0 \\ 0 & 1 & 0 \\ 0 & 0 & 1 \end{bmatrix}$$

$$\mathbf{AA}^{\mathrm{a}} = |\mathbf{A}|\mathbf{I}$$

Hence

$$\frac{\mathbf{AA}^{\mathrm{a}}}{|\mathbf{A}|} = \mathbf{I}$$

Now it has been established that $\mathbf{A}^{-1}\mathbf{A} = \mathbf{I}$. Hence

$$\mathbf{A}^{-1}\mathbf{A} = \frac{\mathbf{AA}^{\mathrm{a}}}{|\mathbf{A}|}$$

or

$$\mathbf{A}^{-1} = \frac{\mathbf{A}^{\mathrm{a}}}{|\mathbf{A}|}$$

The inverse of a matrix is equal to its adjoint matrix divided by its determinant.

Example

Considering the previous matrix

$$\mathbf{A} = \begin{bmatrix} 4 & 3 & 2 \\ 1 & 8 & 4 \\ 6 & 3 & 5 \end{bmatrix}$$

By the Rule of Sarrus $|\mathbf{A}| = 79$. It has been shown that

$$\mathbf{A}^a = \begin{bmatrix} 28 & -9 & -4 \\ 19 & 8 & -14 \\ -45 & 6 & 29 \end{bmatrix}$$

Hence

$$\mathbf{A}^{-1} = \frac{\mathbf{A}^a}{|\mathbf{A}|} = \frac{1}{79} \begin{bmatrix} 28 & -9 & -4 \\ 19 & 8 & -14 \\ -45 & 6 & 29 \end{bmatrix}$$

The purpose of this book is to introduce the reader to stress analysis techniques that can be used in conjunction with computers. Computers are provided with routines to carry out all the usual operations that are performed on matrices, including their inversion, and there is little point in describing all of the methods by which the inverse of a matrix may be found. However there are occasions when it is simpler to obtain an inverse 'by hand' and therefore one other method, applicable to larger matrices, will be described.

The method is one of elimination and is best illustrated by an example.

Suppose that we wish to obtain the inverse of the matrix

$$\begin{bmatrix} 4 & 3 & 2 \\ 1 & 8 & 4 \\ 6 & 3 & 5 \end{bmatrix}$$

The procedure is started by writing down the matrix on the left hand side of a unit matrix:

$$\begin{bmatrix} 4 & 3 & 2 \\ 1 & 8 & 4 \\ 6 & 3 & 5 \end{bmatrix} \begin{bmatrix} 1 & 0 & 0 \\ 0 & 1 & 0 \\ 0 & 0 & 1 \end{bmatrix}$$

The process of finding the inverse consists of gradually changing the left hand matrix into a unit matrix. Every change made to the left hand matrix must be repeated in the right hand matrix. At the end of the operation the right hand matrix will be the inverse of the original matrix.

There are various routes that can be followed to the final result, the point to remember is that one must work towards obtaining zero values above and below the leading diagonal.

$$\begin{bmatrix} 4 & 3 & 2 \\ 1 & 8 & 4 \\ 6 & 3 & 5 \end{bmatrix} \begin{bmatrix} 1 & 0 & 0 \\ 0 & 1 & 0 \\ 0 & 0 & 1 \end{bmatrix}$$

Subtract first row from third row to establish a new third row

$$\begin{bmatrix} 4 & 3 & 2 \\ 1 & 8 & 4 \\ 2 & 0 & 3 \end{bmatrix} \begin{bmatrix} 1 & 0 & 0 \\ 0 & 1 & 0 \\ -1 & 0 & 1 \end{bmatrix}$$

Subtract 3 times the second row from 8 times the first row

$$\begin{bmatrix} 29 & 0 & 4 \\ 1 & 8 & 4 \\ 2 & 0 & 3 \end{bmatrix} \begin{bmatrix} 8 & -3 & 0 \\ 0 & 1 & 0 \\ -1 & 0 & 1 \end{bmatrix}$$

Subtract 4 times the third row from 3 times the first row

$$\begin{bmatrix} 79 & 0 & 0 \\ 1 & 8 & 4 \\ 2 & 0 & 3 \end{bmatrix} \begin{bmatrix} 28 & -9 & -4 \\ 0 & 1 & 0 \\ -1 & 0 & 1 \end{bmatrix}$$

Subtract the third row from 2 times the second row

$$\begin{bmatrix} 79 & 0 & 0 \\ 0 & 16 & 5 \\ 2 & 0 & 3 \end{bmatrix} \begin{bmatrix} 28 & -9 & -4 \\ 1 & 2 & -1 \\ -1 & 0 & 1 \end{bmatrix}$$

Subtract 2 times the first row from 79 times the third row

$$\begin{bmatrix} 79 & 0 & 0 \\ 0 & 16 & 5 \\ 0 & 0 & 237 \end{bmatrix} \begin{bmatrix} 28 & -9 & -4 \\ 1 & 2 & -1 \\ -135 & 18 & 87 \end{bmatrix}$$

Subtract 5 times the third row from 237 times the second row

$$\begin{bmatrix} 79 & 0 & 0 \\ 0 & 3792 & 0 \\ 0 & 0 & 237 \end{bmatrix} \begin{bmatrix} 28 & -9 & -4 \\ 912 & 384 & -672 \\ -135 & 18 & 87 \end{bmatrix}$$

Divide the first row by 79, the second row by 3792, the third row by 237

$$\begin{bmatrix} 1 & 0 & 0 \\ 0 & 1 & 0 \\ 0 & 0 & 1 \end{bmatrix} \begin{bmatrix} \frac{28}{79} & \frac{-9}{79} & \frac{-4}{79} \\ \frac{912}{3792} & \frac{384}{3792} & \frac{-672}{3792} \\ \frac{-135}{237} & \frac{18}{237} & \frac{87}{237} \end{bmatrix}$$

The right hand matrix is the inverse of the original matrix. It can be simplified somewhat to:

$$\frac{1}{79}\begin{bmatrix} 28 & -9 & -4 \\ 19 & 8 & -14 \\ -45 & 6 & 29 \end{bmatrix}$$

Check $\mathbf{AA}^{-1} = \mathbf{I}$

$$\frac{1}{79}\begin{bmatrix} 28 & -9 & -4 \\ 19 & 8 & -14 \\ -45 & 6 & 29 \end{bmatrix} \begin{bmatrix} 4 & 3 & 2 \\ 1 & 8 & 4 \\ 6 & 3 & 5 \end{bmatrix}$$

$$= \frac{1}{79}\begin{bmatrix} 79 & 0 & 0 \\ 0 & 79 & 0 \\ 0 & 0 & 79 \end{bmatrix} = \begin{bmatrix} 1 & 0 & 0 \\ 0 & 1 & 0 \\ 0 & 0 & 1 \end{bmatrix}$$

MULTIPLICATION OF A MATRIX BY A SCALAR QUANTITY

In the previous discussion the multiplication of a matrix by a scalar quantity has been used. A scalar quantity can either pre- or post-multiply a matrix.

i.e.

$$k[\mathbf{A}] = [\mathbf{A}]k$$

$$k[\mathbf{A}][\mathbf{B}] = [\mathbf{A}]k[\mathbf{B}] = [\mathbf{A}][\mathbf{B}]k$$

The scalar quantity is applied to every element in the matrix.

Example

$$k\begin{bmatrix} a_{11} & a_{12} & a_{13} \\ a_{21} & a_{22} & a_{23} \\ a_{31} & a_{32} & a_{33} \end{bmatrix} = \begin{bmatrix} ka_{11} & ka_{12} & ka_{13} \\ ka_{21} & ka_{22} & ka_{23} \\ ka_{31} & ka_{32} & ka_{33} \end{bmatrix}$$

SINGULAR MATRIX

It has been established that the inverse of a matrix [**A**] is obtained from the adjoint matrix [$\mathbf{A}^a$] and the determinant |**A**| by the relationship

$$[\mathbf{A}^{-1}] = \frac{[\mathbf{A}^a]}{|\mathbf{A}|}$$

This relationship can only hold provided that |**A**| is not equal to zero. Hence when $|\mathbf{A}| = 0$ the matrix **A** cannot have an inverse. Such a matrix is said to be singular.

A transformation carried out by a singular matrix is called a singular transformation and is irreversible.

In other words, although the transformation $\mathbf{y} = \mathbf{Ax}$ is possible the reverse transformation $\mathbf{x} = \mathbf{A}^{-1}\mathbf{y}$ cannot be carried out. How does one decide whether or not a matrix is singular? There are some rules that can be of assistance:
A matrix is singular if:

(1) It is rectangular.
(2) It is square but has a row or column of zero elements.
(3) It is square but it has two rows (or two columns) with equal components.
(4) It is square but at least one row (or one column) has a linear relationship to another row (or column).

Rule (3) is really an extension of rule (4). More will be said about linear relationships in the next chapter. With most square matrices it is not generally possible to determine singularity by inspection and the only sure method is to check if the determinant of the matrix is zero.

FURTHER RELATIONSHIPS BETWEEN A MATRIX AND ITS INVERSE

A general rule, that can be verified by the reader with a numerical example, is that if **A**, **B** and **C** are square matrices such that

$$\mathbf{AB} = \mathbf{C}$$

then

$$|\mathbf{A}|\,|\mathbf{B}| = |\mathbf{C}|$$

i.e. the products of the determinants of two square matrices equals the determinant of the product of the matrices.

A special application of this rule is the multiplication of a matrix by its inverse:

$$\mathbf{AA}^{-1} = \mathbf{I}$$

Applying the above rule:

$$|\mathbf{A}|\,|\mathbf{A}^{-1}| = |\mathbf{I}| = 1$$

(The determinant of the unit matrix is unity.)

Hence it can be said that the determinant of a matrix times the determinant of its inverse is always equal to unity.

PARTITIONED MATRICES

In engineering problems it often occurs that a matrix is found to consist of definite groups of elements and, in these cases, it can be convenient to regard these groups as submatrices of the larger matrix.

A similar position arises when the matrix is so large that it may take up considerable storage space in a computer.

In such cases, if we wish, we can divide a matrix into groups of elements. The matrix is then said to be partitioned. The convention is to indicate the subdivisions by dashed lines which, of course, must extend right across the matrix.

Example

$$\mathbf{A} = \left[\begin{array}{cc|cc} a_{11} & a_{12} & a_{13} & a_{14} \\ a_{21} & a_{22} & a_{23} & a_{24} \\ \hdashline a_{31} & a_{32} & a_{33} & a_{34} \end{array}\right]$$

If we denote the submatrices as $\mathbf{A}_{11}$, $\mathbf{A}_{12}$, $\mathbf{A}_{21}$ and $\mathbf{A}_{22}$ the matrix $\mathbf{A}$ can be written as:

$$\mathbf{A} = \begin{bmatrix} \mathbf{A}_{11} & \mathbf{A}_{12} \\ \mathbf{A}_{21} & \mathbf{A}_{22} \end{bmatrix}$$

where

$$\mathbf{A}_{11} = \begin{bmatrix} \mathbf{a}_{11} & \mathbf{a}_{12} \\ \mathbf{a}_{21} & a_{22} \end{bmatrix} \quad \mathbf{A}_{12} = \begin{bmatrix} a_{13} & a_{14} \\ a_{23} & a_{24} \end{bmatrix}$$

$$\mathbf{A}_{21} = [a_{31} \quad a_{32}] \quad \mathbf{A}_{22} = [a_{33} \quad a_{34}]$$

The rules for partioned matrices are exactly as for normal matrices. They can be added, subtracted and multiplied in the normal fashion always provided that the partitioning has been undertaken in a suitable manner.

Example

Determine the inverse of

$$\mathbf{A} = \begin{bmatrix} 1 & 0 & 0 & 0 & 0 & 0 \\ 1 & 4 & 6 & 0 & 0 & 0 \\ 1 & 3 & 2 & 0 & 0 & 0 \\ 0 & 0 & 0 & 1 & 0 & 0 \\ 0 & 0 & 0 & 1 & 4 & 6 \\ 0 & 0 & 0 & 1 & 3 & 2 \end{bmatrix}$$

If we partition the matrix thus:

$$\left[\begin{array}{ccc|ccc} 1 & 0 & 0 & 0 & 0 & 0 \\ 1 & 4 & 6 & 0 & 0 & 0 \\ 1 & 3 & 2 & 0 & 0 & 0 \\ \hline 0 & 0 & 0 & 1 & 0 & 0 \\ 0 & 0 & 0 & 1 & 4 & 6 \\ 0 & 0 & 0 & 1 & 3 & 2 \end{array}\right]$$

Then

$$\mathbf{A}_{11} = \mathbf{A}_{22} = \begin{bmatrix} 1 & 0 & 0 \\ 1 & 4 & 6 \\ 1 & 3 & 2 \end{bmatrix} \quad \text{and} \quad \mathbf{A}_{12} = \mathbf{A}_{21} = 0$$

Hence

$$\mathbf{A} = \begin{bmatrix} \mathbf{A}_{11} & 0 \\ 0 & \mathbf{A}_{22} \end{bmatrix}$$

Now

$$\mathbf{A}\mathbf{A}^{-1} = \begin{bmatrix} 1 & 0 \\ 0 & 1 \end{bmatrix}$$

Hence

$$\mathbf{A}^{-1} = \begin{bmatrix} \mathbf{A}_{11}{}^{-1} & 0 \\ 0 & \mathbf{A}_{22}{}^{-1} \end{bmatrix}$$

We can determine the inverse of $\mathbf{A}_{11}$ in various ways. We will use the adjoint matrix here.

$$\mathbf{A}_c = \begin{bmatrix} 10 & 4 & -1 \\ 0 & 2 & -3 \\ 0 & -6 & 4 \end{bmatrix}$$

$$\mathbf{A}_a = \begin{bmatrix} -10 & 0 & 0 \\ 4 & 2 & -6 \\ -1 & -3 & 4 \end{bmatrix}$$

Now

$$|\mathbf{A}_{11}| = -10$$

Hence

$$\mathbf{A}_{11}{}^{-1} = \frac{\mathbf{A}_{11a}}{|\mathbf{A}|} = \begin{bmatrix} 1 & 0 & 0 \\ -0{\cdot}4 & -0{\cdot}2 & 0{\cdot}6 \\ 0{\cdot}1 & 0{\cdot}3 & -0{\cdot}4 \end{bmatrix}$$

Hence

$$\mathbf{A}^{-1} = \begin{bmatrix} 1 & 0 & 0 & 0 & 0 & 0 \\ -0{\cdot}4 & -0{\cdot}2 & 0{\cdot}6 & 0 & 0 & 0 \\ 0{\cdot}1 & 0{\cdot}3 & -0{\cdot}4 & 0 & 0 & 0 \\ 0 & 0 & 0 & 1 & 0 & 0 \\ 0 & 0 & 0 & -0{\cdot}4 & -0{\cdot}2 & 0{\cdot}6 \\ 0 & 0 & 0 & 0{\cdot}1 & 0{\cdot}3 & -0{\cdot}4 \end{bmatrix}$$

which may be checked by multiplying out $\mathbf{A}\mathbf{A}^{-1}$ to yield a unit matrix.

COMMUTATIVE, ASSOCIATIVE AND DISTRIBUTIVE LAWS OF MATRIX ALGEBRA

In order to define these terms let us consider some algebraic expressions first.

THE COMMUTATIVE LAW

$$ab = ba$$

The symbols can change places, or commute, without altering the value of the expression.

THE ASSOCIATIVE LAW

$$a + b + c = (a + b) + c = a + (b + c)$$

b can be associated with either *a* or *c* without altering the value of the expression.

THE DISTRIBUTIVE LAW

$$ab + bc = b(a + c)$$

The various terms can be redistributed without altering the value of the expression.

In matrix algebra the laws (which have been gradually established throughout this chapter) are:

FOR ADDITION OF MATRICES

Matrices obey both the commutative and the associative laws:

$$\mathbf{X} + \mathbf{Y} = \mathbf{Y} + \mathbf{X}$$

$$\mathbf{X} + \mathbf{Y} + \mathbf{Z} = (\mathbf{X} + \mathbf{Y}) + \mathbf{Z} = \mathbf{X} + (\mathbf{Y} + \mathbf{Z})$$

FOR MULTIPLICATION OF MATRICES

Multiplication is not commutative:

$$\mathbf{AB} \neq \mathbf{BA}$$

Multiplication is associative:

$$\mathbf{XYZ} = (\mathbf{XY})\mathbf{Z} = \mathbf{X}(\mathbf{YZ})$$

Multiplication is distributive:

$$\mathbf{X}(\mathbf{Y} + \mathbf{Z}) = \mathbf{XY} + \mathbf{XZ}$$

and

$$(\mathbf{Y} + \mathbf{Z})\mathbf{X} = \mathbf{YX} + \mathbf{ZX}$$

Note. $(\mathbf{Y} + \mathbf{Z})\mathbf{X} \neq \mathbf{XY} + \mathbf{XZ}$ as multiplication is not commutative.

4

Transformations and Eigenvalues

SIMULTANEOUS EQUATIONS

SIMULTANEOUS equations are often encountered in engineering. In a large amount of stress analysis work linearity is assumed resulting in simultaneous equations of the first degree.

NONHOMOGENEOUS EQUATIONS

Let us consider a set of first degree equations which relate two variables, x and y as follows:

$$
\begin{array}{l}
y_1 = a_{11}x_1 + a_{12}x_2 + a_{13}x_3 \ldots + a_{1n}x_n \\
y_2 = a_{21}x_1 + a_{22}x_2 + a_{23}x_3 \ldots + a_{2n}x_n \\
\cdot \quad \cdot \quad \cdot \quad \cdot \quad \ldots \quad \cdot \\
\cdot \quad \cdot \quad \cdot \quad \cdot \quad \ldots \quad \cdot \\
\cdot \quad \cdot \quad \cdot \quad \cdot \quad \ldots \quad \cdot \\
y_n = a_{n1}x_1 + a_{n2}x_2 + a_{n3}x_3 \ldots + a_{nn}x_n
\end{array}
$$

These relationships can be expressed in matrix form

$$\mathbf{y} = \mathbf{A}\mathbf{x}$$

where

$$
\mathbf{y} = \begin{bmatrix} y_1 \\ y_2 \\ y_3 \\ \cdot \\ \cdot \\ \cdot \\ y_n \end{bmatrix} \quad
\mathbf{A} = \begin{bmatrix} a_{11} & a_{12} & a_{13} \ldots a_{1n} \\ a_{21} & a_{22} & a_{23} \ldots a_{2n} \\ \cdot & \cdot & \cdot \quad \ldots\ldots \\ \cdot & \cdot & \cdot \quad \ldots\ldots \\ \cdot & \cdot & \cdot \quad \ldots\ldots \\ a_{n1} & a_{n2} & a_{n3} \ldots a_{nn} \end{bmatrix} \quad
\mathbf{x} = \begin{bmatrix} x_1 \\ x_2 \\ x_3 \\ \cdot \\ \cdot \\ \cdot \\ x_n \end{bmatrix}
$$

Equations of this form, when the vector **y** is not a null vector, are known as nonhomogeneous simultaneous equations. There are various methods available for their solution.

(a) BY ELIMINATION

This is a widely used method and consists of working from one row to the next gradually expressing all the unknowns in terms of one unknown a process which eventually leads to a solution.

Example

Determine x_1, x_2 and x_3 given the following relationships

$$2x_1 + 3x_2 + x_3 = 15 \tag{1}$$

$$x_1 - x_2 + 3x_3 = 10 \tag{2}$$

$$x_1 + 2x_2 - x_3 = 3 \tag{3}$$

From (1)

$$x_1 = \frac{15 - 3x_2 - x_3}{2} = 10 - 3x_3 + x_2 \text{ (From (2))}$$

$$\therefore \quad 15 - 3x_2 - x_3 = 20 - 6x_3 + 2x_2$$

Hence

$$x_2 = x_3 - 1$$

And

$$x_1 = 10 - 3x_3 + x_2 = 10 - 3x_3 + x_3 - 1 = 9 - 2x_3$$

Substituting in (3)

$$9 - 2x_3 + 2x_3 - 2 - x_3 = 3$$

Hence

$$x_3 = 4$$

And

$$x_2 = 3 \qquad x_1 = 1.$$

(b) USING THE INVERSE MATRIX

If we write the three equations of the example in matrix form:

$$\begin{bmatrix} 2 & 3 & 1 \\ 1 & -1 & 3 \\ 1 & 2 & -1 \end{bmatrix} \begin{bmatrix} x_1 \\ x_2 \\ x_3 \end{bmatrix} = \begin{bmatrix} 15 \\ 10 \\ 3 \end{bmatrix}$$

This is an equation of the form $\mathbf{Ax} = \mathbf{y}$. Matrix $\mathbf{A}$ is square and its inverse can be obtained by one of the methods described previously to give the expression $\mathbf{x} = \mathbf{A}^{-1}\mathbf{y}$

$$\mathbf{A}^{-1} = \frac{1}{5}\begin{bmatrix} -5 & 5 & 10 \\ 4 & -3 & -5 \\ 3 & -1 & -5 \end{bmatrix}$$

Hence

$$\begin{bmatrix} x_1 \\ x_2 \\ x_3 \end{bmatrix} = \frac{1}{5}\begin{bmatrix} -5 & 5 & 10 \\ 4 & -3 & -5 \\ 3 & -1 & -5 \end{bmatrix}\begin{bmatrix} 15 \\ 10 \\ 3 \end{bmatrix} = \begin{bmatrix} 1 \\ 3 \\ 4 \end{bmatrix}$$

Hence

$$x_1 = 1, \quad x_2 = 3, \quad x_3 = 4$$

(c) BY DETERMINANTS (Cramer's rule)

For those interested the proof of this rule is given in the appendix. Cramer's rule states that if a set of unknown variables, $x_1, x_2, \ldots, x_n$ is related to another set of variables, $y_1, y_2, \ldots, y_n$ then each of the unknowns is equal to the ratio of two determinants.

To evaluate the general unknown x_i the denominator of x_i is the determinant of the coefficients of the x terms and the numerator is the same determinant except that the ith column is replaced by the terms $y_1, y_2, \ldots, y_n$.

Example

Applying Cramer's rule to the previous example:

$$x_1 = \frac{|\mathbf{A}| \text{ with first column replaced by } y \text{ terms}}{|\mathbf{A}|}$$

$$= \frac{\text{Determinant of coefficients with first column altered}}{\text{Determinant of coefficients}}$$

$$\therefore \quad x_1 = \frac{\begin{vmatrix} 15 & 3 & 1 \\ 10 & -1 & 3 \\ 3 & 2 & -1 \end{vmatrix}}{\begin{vmatrix} 2 & 3 & 1 \\ 1 & -1 & 3 \\ 1 & 2 & -1 \end{vmatrix}} = \frac{5}{5} = 1$$

$$x_2 = \frac{\begin{vmatrix} 2 & 15 & 1 \\ 1 & 10 & 3 \\ 1 & 3 & -1 \end{vmatrix}}{5} = \frac{15}{5} = 3$$

$$x_3 = \frac{\begin{vmatrix} 2 & 3 & 15 \\ 1 & -1 & 10 \\ 1 & 2 & 3 \end{vmatrix}}{5} = \frac{20}{5} = 4$$

(d) THE GAUSS–SEIDEL METHOD

This is an iterative process which can become quite lengthy but is useful in computer programmes. A different unknown is transferred to the right hand side of each equation so that the equations are written in the form:

$$a_{11}x_1 = y_1 - (a_{12}x_2 + a_{13}x_3 + \ldots + a_{1n}x_n)$$
$$a_{22}x_2 = y_2 - (a_{21}x_1 + a_{23}x_3 + \ldots + a_{2n}x_n)$$
$$\begin{matrix} . & . & . & . & \ldots & . \\ . & . & . & . & \ldots & . \\ . & . & . & . & \ldots & . \end{matrix}$$
$$a_{nn}x_n = y_n - (a_{n1}x_1 + a_{n2}x_2 + \ldots + a_{n(n-1)}x_{n(n-1)})$$

Considering the first of these equations: A series of values for x_2, $x_3, \ldots, x_n$ are estimated and inserted in the equation to give some value for x_1. This value for x_1, together with the assumed values for $x_2, x_3, \ldots, x_n$ are now inserted into the second equation to determine a value for x_2. The derived values for x_1 and x_2, together with the assumed values for x_3 to x_n are then substituted into the third equation to give a value for x_3. The process is repeated until a complete set of values for x_1 to x_n has been obtained.

The set of derived values is now used to go through the same procedure again leading to a new set of values for x_1 to x_n. This iterative procedure is continued until the derived values are the same as those obtained with the previous iteration.

Example

Solve the equations

$$2x_1 + x_2 = 7$$
$$3x_1 - x_2 = 3$$

Rewrite the equations as:

$$x_2 = 7 - 2x_1 \tag{1}$$

$$x_1 = \frac{x_2}{3} + 1 \tag{2}$$

As a starting point put $x_1 = 0$ in Equation (1) then $x_2 = 7$. Put $x_2 = 7$ in Equation (2) then $x_1 = 3{\cdot}33$.

Hence, at the end of the first iteration $x_1 = 3{\cdot}33$ and $x_2 = 7$. Substituting $x_1 = 3{\cdot}33$ in Equation (1) gives $x_2 = 0{\cdot}33$ and, substituting $x_2 = 0{\cdot}33$ in Equation (2) gives $x_1 = 1{\cdot}11$.

The process is continued until the values are the same as those obtained previously:

x_1	x_2
0	7·
3·33	0·33
1·11	4·78
2·59	1·82
1·61	3·78
2·26	2·87
1·96	3·08
.	.
.	.
.	.
2·00	3·00

HOMOGENEOUS EQUATIONS

When **y**, in the expression **y** = **Ax** has all its components equal to zero, i.e. when **y** is a null vector, the equations are known as homogeneous simultaneous equations.

The general form of these equations is:

$$\begin{array}{l}
a_{11}x_1 + a_{12}x_2 + a_{13}x_3 + \ldots + a_{1n}x_n = 0 \\
a_{21}x_1 + a_{22}x_2 + a_{23}x_3 + \ldots + a_{2n}x_n = 0 \\
\quad\cdot \qquad\quad \cdot \qquad\quad \cdot \qquad \ldots \quad \cdot \qquad \cdot \\
\quad\cdot \qquad\quad \cdot \qquad\quad \cdot \qquad \ldots \quad \cdot \qquad \cdot \\
\quad\cdot \qquad\quad \cdot \qquad\quad \cdot \qquad \ldots \quad \cdot \qquad \cdot \\
a_{n1}x_1 + a_{n2}x_2 + a_{n3}x_3 + \ldots + a_{nn}x_n = 0
\end{array}$$

In matrix form

$$\mathbf{Ax} = 0$$

The obvious solution, and one that is correct, is that all the components of **x** are zero, i.e. **x** is a null vector. This solution is often referred to as the trivial solution. Any other solutions are known as nontrivial solutions. It is the nontrivial solutions that are generally required.

If we apply Cramer's rule then, for any term x_i,

$$x_i = \frac{\text{Determinant of } \mathbf{A} \text{ with column } i \text{ changed to elements of } \mathbf{y}}{\text{Determinant of } \mathbf{A}}$$

Now vector **y** is a null vector. Hence the numerator of x_i consists of a determinant which has a column of zero elements. Such a determinant is always zero (See Chapter 1). Hence

$$x_1 = \frac{\mathbf{O}}{|\mathbf{A}|}, \quad x_2 = \frac{\mathbf{O}}{|\mathbf{A}|}, \ldots, x_n = \frac{\mathbf{O}}{|\mathbf{A}|}$$

Hence vector

$$\mathbf{x} = \frac{\mathbf{O}}{|\mathbf{A}|}$$

where **O** is a null vector. Hence

$$|\mathbf{A}|\mathbf{x} = \mathbf{O}$$

If determinant **A** has any value other than zero then, for the above expression to be true, **x** must be a null vector (which leads to the trivial solution). However if $|\mathbf{A}|$ is itself zero then **x** can have a value. Hence, in the solution of homogeneous simultaneous equations, the matrix **A** must be singular if nontrivial solutions are to exist.

Example

Determine the values of x_1 and x_2 given the following relationship:

$$3x_1 + 2x_2 = 0$$
$$x_1 - 5x_2 = 0$$

The matrix of the coefficients,

$$\mathbf{A} = \begin{bmatrix} 3 & 2 \\ 1 & -5 \end{bmatrix}$$

$$|\mathbf{A}| = -3 \times 5 - 2 \times 1 = -17$$

As its determinant is not zero matrix **A** is not singular, hence the only solution that exists is the trivial one ($x_1 = x_2 = 0$).

Example

Determine x_1 and x_2 given that:

$$3x_1 + 2x_2 = 0$$

$$6x_1 + 4x_2 = 0$$

$$\mathbf{A} = \begin{bmatrix} 3 & 2 \\ 6 & 4 \end{bmatrix} \quad |\mathbf{A}| = 3 \times 4 - 2 \times 6 = 0$$

Nontrivial solutions exist. By elimination

$$x_1 = -\tfrac{2}{3}x_2$$

Hence, if we assume a set of values for x_2, we can obtain the corresponding values of x_1.

The general solution is

$$x = \beta \begin{bmatrix} -\frac{2}{3} \\ 1 \end{bmatrix}$$

Where β is an arbitrary constant.

The technique applies to any number of homogeneous simultaneous equations.

LINEAR DEPENDENCE

A set of variables, $x_1, x_2, x_3 \ldots, x_n$ are said to be linearly dependent when they satisfy an equation of the form:

$$\alpha_1 x_1 + \alpha_2 x_2 + \alpha_3 x_3 + \ldots + \alpha_n x_n = 0$$

where at least one of the coefficients α is not equal to zero.

If this equation is only possible when all the coefficients are zero the quantities are said to be linearly independent.

Example

Check whether or not the quantities $(3x + y)$ and $(6x + 2y)$ are linearly dependent.

Express the quantities in an equation of the form:

$$\alpha_1(3x + y) + \alpha_2(6x + 2y) = 0$$

It can be seen by inspection that this equation will apply when $\alpha_1 = -2$ and $\alpha_2 = 1$. Hence the two quantities are linearly dependent.

Example

Check the linear dependence of the following quantities:

$$(-x + 2y + 3z); \quad (x - 4y - 13z); \quad (-3x + 5y + 4z)$$

Express in the form:

$$\alpha_1(-x + 2y + 3z) + \alpha_2(x - 4y - 13z) + \alpha_3(-3x + 5y + 4z) = 0$$

i.e.

$$x(-\alpha_1 + \alpha_2 - 3\alpha_3) = 0$$
$$y(2\alpha_1 - 4\alpha_2 + 5\alpha_3) = 0$$
$$z(3\alpha_1 - 13\alpha_2 + 4\alpha_3) = 0$$

Hence we have three simultaneous equations:

$$-\alpha_1 + \alpha_2 - 3\alpha_3 = 0 \qquad \text{(i)}$$
$$2\alpha_1 - 4\alpha_2 + 5\alpha_3 = 0 \qquad \text{(ii)}$$
$$3\alpha_1 - 13\alpha_2 + 4\alpha_3 = 0 \qquad \text{(iii)}$$

From (i)

$$\alpha_1 = \alpha_2 - 3\alpha_3$$

Substituting in (ii):

$$2\alpha_2 - 6\alpha_3 - 4\alpha_2 + 5\alpha_3 = 0$$

$$\therefore \quad \alpha_2 = -\frac{\alpha_3}{2}$$

Hence

$$\alpha_1 = -\frac{\alpha_3}{2} - 3\alpha_3 = -\frac{7}{2}\alpha_3$$

Substituting for α_1 and α_2 in (iii)

$$-\frac{3 \times 7}{2}\alpha_3 + \frac{13\alpha_3}{2} + 4\alpha_3 = 0$$

$$-10{\cdot}5\alpha_3 + 6{\cdot}5\alpha_3 + 4\alpha_3 = 0$$

This is possible for any value of α_3; hence the quantities are linearly dependent.

Example

Check whether or not the following vectors are linearly dependent.

$$\begin{bmatrix} 2 \\ 4 \\ 3 \end{bmatrix} \quad \begin{bmatrix} 0 \\ 6 \\ 4 \end{bmatrix} \quad \begin{bmatrix} 8 \\ 4 \\ 6 \end{bmatrix}$$

The expression

$$a\begin{bmatrix}2\\4\\3\end{bmatrix} + b\begin{bmatrix}0\\6\\4\end{bmatrix} + c\begin{bmatrix}8\\4\\6\end{bmatrix} = 0$$

leads to the three simultaneous equations:

$$2a + 8c = 0 \qquad \text{(i)}$$

$$4a + 6b + 4c = 0 \qquad \text{(ii)}$$

$$3a + 4b + 6c = 0 \qquad \text{(iii)}$$

From (i) $a = -4c$, substituting in (ii)

$$-16c + 6b + 4c = 0$$

Hence $b = 2c$. Substituting for a and b in (iii):

$$-12c + 8c + 6c = 0$$

This is only possible if $c = 0$ hence the three vectors are linearly independent.

A more elegant solution is that, as the three simultaneous equations are homogeneous the determinant of the coefficients must be singular if there are to be nontrivial solutions:

$$\begin{vmatrix}2 & 0 & 8\\4 & 6 & 4\\3 & 4 & 6\end{vmatrix} = 24$$

The matrix is not singular and the only solution is $a = b = c = 0$, and the vectors are linearly independent. The reader should use this method to check the previous example.

EIGENVALUES AND EIGENVECTORS

In many engineering applications, particularly in transformations (discussed later) one encounters a set of simultaneous equations in the form:

$$\mathbf{Ax} = \lambda\mathbf{x}.$$

Where

$\mathbf{A}$ = a square matrix

$\mathbf{x}$ = a column vector

λ = a scalar quantity

Example

Consider the two simultaneous equations:

$$5x_1 - 2x_2 = \lambda x_1 \qquad (1)$$

$$x_1 + 2x_2 = \lambda x_2 \qquad (2)$$

These two equations can be expressed as homogeneous simultaneous equations:

$$(5 - \lambda)x_1 - 2x_2 = 0$$

$$x_1 + (2 - \lambda)x_2 = 0$$

The coefficient matrix is:

$$\begin{bmatrix} 5 - \lambda & -2 \\ 1 & 2 - \lambda \end{bmatrix}$$

For nontrivial solutions to be possible this matrix must be singular i.e.

$$(5 - \lambda)(2 - \lambda) + 2 = 0$$

$$\lambda^2 - 7\lambda + 10 + 2 = 0$$

$$\therefore \quad \lambda = 3 \quad \text{or} \quad \lambda = 4$$

If we look more closely at the coefficient matrix

$$\begin{bmatrix} 5 - \lambda & -2 \\ 1 & 2 - \lambda \end{bmatrix}$$

we see that it can be written as:

$$\begin{bmatrix} 5 & -2 \\ 1 & 2 \end{bmatrix} + \begin{bmatrix} -\lambda & 0 \\ 0 & -\lambda \end{bmatrix} \quad \text{or} \quad \begin{bmatrix} 5 & -2 \\ 1 & 2 \end{bmatrix} - \begin{bmatrix} 1 & 0 \\ 0 & 1 \end{bmatrix} \lambda$$

The example has shown that the expression $\mathbf{Ax} = \lambda\mathbf{x}$ can be written in the form:

$$(\mathbf{A} - \mathbf{I}\lambda)\mathbf{x} = \mathbf{0}$$

where $\mathbf{I}$ is a unit matrix of the same order as matrix $\mathbf{A}$ and $\mathbf{0}$ is a null matrix.

$(\mathbf{A} - \mathbf{I}\lambda)$ is the coefficient matrix and, as is the case with homogeneous equations, its determinant must be zero if nontrivial solutions are to exist. The equation $|\mathbf{A} - \mathbf{I}\lambda| = 0$ is known as the characteristic equation of matrix $\mathbf{A}$. The values of λ that satisfy the equation are known as the eigenvalues of the matrix $\mathbf{A}$.

SPECTRAL MATRIX

This particular matrix is important in stress analysis. It is simply a diagonal matrix in which the elements of the leading diagonal are the eigenvalues. It is generally given the symbol **S**.
i.e.

$$\mathbf{S} = \begin{bmatrix} \lambda_1 & 0 & 0 \ldots & 0 \\ 0 & \lambda_2 & 0 \ldots & 0 \\ 0 & . & \ldots & 0 \\ 0 & . & \ldots & \lambda_n \end{bmatrix}$$

In our example

$$\mathbf{S} = \begin{bmatrix} 3 & 0 \\ 0 & 4 \end{bmatrix}$$

EIGENVECTORS

Having obtained the eigenvalues of a matrix the corresponding eigenvectors can be obtained. In other words, having found λ, then the corresponding value of **x** in the expression $(\mathbf{A} - \mathbf{I}\lambda)x = 0$ can be found.

The procedure is to solve the homogeneous equations formed from $(\mathbf{A} - \mathbf{I}\lambda)\mathbf{x} = \mathbf{O}$. For each value of λ there will be a corresponding value of **x**, the eigenvector.

In the two dimensional example of eigenvalues the matrix **A** was

$$\mathbf{A} = \begin{bmatrix} 5 & -2 \\ 1 & 2 \end{bmatrix} \quad \text{and} \quad \mathbf{I}\lambda = \begin{bmatrix} \lambda & 0 \\ 0 & \lambda \end{bmatrix}$$

Hence

$$(\mathbf{A} - \mathbf{I}\lambda) = \begin{bmatrix} 5 - \lambda & -2 \\ 1 & 2 - \lambda \end{bmatrix}$$

Hence the expression $(\mathbf{A} - \mathbf{I}\lambda)\mathbf{x} = \mathbf{O}$ can be written as:

$$\begin{bmatrix} 5 - \lambda & -2 \\ 1 & 2 - \lambda \end{bmatrix} \begin{bmatrix} x_1 \\ x_2 \end{bmatrix} = \begin{bmatrix} 0 \\ 0 \end{bmatrix}$$

and, substituting for the first value $\lambda = 3$

$$\begin{bmatrix} 5 - 3 & -2 \\ 1 & 2 - 3 \end{bmatrix} \begin{bmatrix} x_1 \\ x_2 \end{bmatrix} = \begin{bmatrix} 0 \\ 0 \end{bmatrix}$$

Hence

$$2x_1 - 2x_2 = 0$$

$$x_1 - x_2 = 0$$

From these two equations we see that $x_1 = x_2$. Hence we can establish the first eigenvector

$$x = \begin{bmatrix} x_1 \\ x_2 \end{bmatrix}$$

if we assume a value for either x_1 or x_2.

The general solution is

$$x = \beta_1 \begin{bmatrix} 1 \\ 1 \end{bmatrix}$$

where β_1 is an arbitrary constant.

The second eigenvector is obtained by substituting $\lambda = 4$ in the equation:

$$\begin{bmatrix} 5-4 & -2 \\ 1 & 2-4 \end{bmatrix} \begin{bmatrix} x_1 \\ x_2 \end{bmatrix} = \begin{bmatrix} 0 \\ 0 \end{bmatrix}$$

Hence

$$x_1 - 2x_2 = 0$$

$$x_1 - 2x_2 = 0$$

i.e.

$$x_1 = 2x_2$$

The general solution for this vector is

$$\beta_2 \begin{bmatrix} 1 \\ \frac{1}{2} \end{bmatrix},$$

where β_2 is an arbitrary constant.

In order to have only one set of components for each eigenvector one of two procedures may be used.

The first method is to put the largest component equal to unity and evaluate the other components accordingly.

By this method the eigenvectors in the example are:

$$\text{First eigenvector} = \begin{bmatrix} 1 \\ 1 \end{bmatrix} \quad \text{Second eigenvector} \begin{bmatrix} 1 \\ \frac{1}{2} \end{bmatrix}$$

The other method is to normalise the eigenvectors—this is the method used in stress analysis. From Chapter 2 it will be remembered that a normalised vector is of unit length and is obtained by dividing each component of the vector by its length. The first step is to make one of the components (any one) equal to unity. Evaluate the other components, determine the length and divide each component by the length. By this method the eigenvectors in the example are:

$$\text{First eigenvector } \mathbf{x} = \begin{bmatrix} 1 \\ 1 \end{bmatrix} \quad \text{Length} = \sqrt{(1^2 + 1^2)} = \sqrt{2}$$

$$\text{First normalised eigenvector} = \begin{bmatrix} \dfrac{1}{\sqrt{2}} \\ \dfrac{1}{\sqrt{2}} \end{bmatrix}$$

Similarly the second normalised vector is

$$\begin{bmatrix} \dfrac{2}{\sqrt{5}} \\ \dfrac{1}{\sqrt{5}} \end{bmatrix}$$

MODAL MATRIX

A square $n \times n$ matrix obviously has n eigenvectors. These eigenvectors, which will consist of n elements, can be written as columns of a square matrix, called a modal matrix. (An eigenvector is sometimes referred to as a modal column.) The modal matrix is given the symbol **M**. If, in the previous example β_1 and β_2 are put equal to 1 then

$$\mathbf{M} = \begin{bmatrix} 1 & 1 \\ 1 & \frac{1}{2} \end{bmatrix}$$

Similarly the modal matrix of the normalised vectors is:

$$\mathbf{N} = \begin{bmatrix} \dfrac{1}{\sqrt{2}} & \dfrac{2}{\sqrt{5}} \\ \dfrac{1}{\sqrt{2}} & \dfrac{1}{\sqrt{5}} \end{bmatrix}$$

Note that the symbol **N** is used instead of **M** when the vectors have been normalised.

Example

Consider the matrix

$$\mathbf{A} = \begin{bmatrix} 12 & -5 & 2 \\ -5 & 9 & -4 \\ 2 & -4 & 7 \end{bmatrix}$$

and let us evaluate all the terms that have been discussed in this chapter so far. The characteristic equation of the matrix is:

$$\begin{vmatrix} 12 - \lambda & -5 & 2 \\ -5 & 9 - \lambda & -4 \\ 2 & -4 & 7 - \lambda \end{vmatrix} = 0$$

If expanded the resulting cubic expression is:

$$\lambda^3 - 28\lambda^2 + 210\lambda - 433 = 0$$

This expression factorises to

$$(\lambda - 17{\cdot}32)(\lambda - 7{\cdot}22)(\lambda - 3{\cdot}46) = 0$$

(A method for determining the roots of a cubic equation is given on page 202.)

Hence the eigenvalues of matrix **A** are:

$$\lambda_1 = 17{\cdot}32$$
$$\lambda_2 = 7{\cdot}22$$
$$\lambda_3 = 3{\cdot}46$$

The spectral matrix is therefore:

$$\mathbf{S} = \begin{bmatrix} 17{\cdot}32 & 0 & 0 \\ 0 & 7{\cdot}22 & 0 \\ 0 & 0 & 3{\cdot}46 \end{bmatrix}$$

Let us now evaluate the eigenvectors.

Considering $\lambda_1 = 17{\cdot}32$, $(\mathbf{A} - \mathbf{I}\lambda)\mathbf{x} = \mathbf{O}$ can be expressed as

$$\begin{bmatrix} -5{\cdot}32 & -5 & 2 \\ -5 & -8{\cdot}32 & -4 \\ 2 & -4 & -10{\cdot}32 \end{bmatrix} \begin{bmatrix} x_1 \\ x_2 \\ x_3 \end{bmatrix} = \begin{bmatrix} 0 \\ 0 \\ 0 \end{bmatrix}$$

This is a set of homogeneous equations and to obtain nontrivial solutions the matrix $(\mathbf{A} - \mathbf{I}\lambda)$ must be singular. This must be the case here as we are using the three values of λ which satisfy this condition.

$$\therefore \quad -5{\cdot}32x_1 - 5x_2 + 2x_3 = 0 \qquad (1)$$

$$-5x_1 - 8{\cdot}32x_2 - 4x_3 = 0 \qquad (2)$$

(The third equation is not required.)

Using the method of elimination: From Equation (1)

$$x_1 = \frac{-5x_2 + 2x_3}{5{\cdot}32} = -0{\cdot}94x_2 + 0{\cdot}376x_3$$

From Equation (2)

$$5(-0{\cdot}94x_2 + 0{\cdot}376x_3) = -8{\cdot}32x_2 - 4x_3$$

$$\therefore \quad -0{\cdot}94x_2 + 0{\cdot}376x_3 = 1{\cdot}664x_2 - 0{\cdot}8x_3$$

$$\therefore \quad x_2 = -1{\cdot}625x_3.$$

Hence

$$x_1 = 0{\cdot}94 \times 1{\cdot}625x_3 + 0{\cdot}376x_3 = 1{\cdot}903x_3$$

As x_1 is the largest put $x_1 = 1{\cdot}0$. Then

$$x_3 = \frac{1}{1{\cdot}903} = 0{\cdot}524$$

and

$$x_2 = -1{\cdot}625 \times 0{\cdot}524 = -0{\cdot}852.$$

Hence the general solution for the first eigenvector is

$$\beta_1 \begin{bmatrix} 1{\cdot}0 \\ -0{\cdot}852 \\ 0{\cdot}524 \end{bmatrix}$$

If $\beta_1 = 1$ the modal column is

$$\begin{bmatrix} 1{\cdot}0 \\ -0{\cdot}852 \\ 0{\cdot}524 \end{bmatrix}$$

Note. Mistakes are all too easily made in the solution of equations such as those just solved. It is well worth the trouble to check the values obtained by inserting them in the third equation. In the example the third equation is: $2x_1 - 4x_2 - 10{\cdot}32x_3 = 0$.

$$2 \times 1 + 4 \times 0{\cdot}852 - 10{\cdot}32 \times 0{\cdot}524 = 0.$$

The normalised eigenvector can also be found. Taking the modal column: length $= \sqrt{(1^2 + 0{\cdot}852^2 + 0{\cdot}524^2)} = \sqrt{2} = 1{\cdot}414$.

Hence normalised eigenvector is

$$\begin{bmatrix} \dfrac{1}{1{\cdot}414} \\ \dfrac{-0{\cdot}852}{1{\cdot}414} \\ \dfrac{0{\cdot}524}{1{\cdot}414} \end{bmatrix} = \begin{bmatrix} 0{\cdot}708 \\ -0{\cdot}603 \\ 0{\cdot}37 \end{bmatrix}$$

These values can also be checked by insertion in any one of the three equations. The reader may like to solve for the other two eigenvectors. When this has been accomplished the modal matrix is found to be:

$$\mathbf{M} = \begin{bmatrix} 1{\cdot}0 & -0{\cdot}996 & 0{\cdot}37 \\ -0{\cdot}852 & -0{\cdot}554 & 1{\cdot}0 \\ 0{\cdot}524 & 1{\cdot}0 & 0{\cdot}927 \end{bmatrix}$$

And the normalised modal matrix is:

$$\mathbf{N} = \begin{bmatrix} 0{\cdot}708 & -0{\cdot}658 & 0{\cdot}262 \\ -0{\cdot}603 & -0{\cdot}366 & 0{\cdot}71 \\ 0{\cdot}37 & 0{\cdot}660 & 0{\cdot}657 \end{bmatrix}$$

MATRIX DIAGONALISATION

As explained, the modal matrix, $\mathbf{M} = [\mathbf{x}_1, \mathbf{x}_2, \ldots, \mathbf{x}_n]$ consists of the components of the eigenvectors of matrix **A**, arranged as columns. These eigenvectors, $\mathbf{x}_1, \mathbf{x}_2, \ldots, \mathbf{x}_n$ satisfy the general expression

$$\mathbf{Ax} = \lambda\mathbf{x}$$

or

$$\begin{aligned} \mathbf{Ax}_1 &= \lambda_1\mathbf{x}_1 \\ \mathbf{Ax}_2 &= \lambda_2\mathbf{x}_2 \\ \cdot \quad & \quad \cdot \\ \cdot \quad & \quad \cdot \\ \cdot \quad & \quad \cdot \\ \mathbf{Ax}_n & \quad \lambda_n\mathbf{x}_n \end{aligned}$$

If we premultiply the modal matrix by the matrix **A** we obtain:

$$\begin{aligned} \mathbf{AM} &= \mathbf{A}[\mathbf{x}_1, \mathbf{x}_2, \ldots, \mathbf{x}_n] \\ &= [\mathbf{Ax}_1\ \mathbf{Ax}_2, \ldots, \mathbf{Ax}_n] \\ &= [\lambda_1\mathbf{x}_1\ \lambda_2\mathbf{x}_2 \ldots, \lambda_n\mathbf{x}_n] \end{aligned}$$

This latter expression is a modified form of the modal matrix **M**. Each column of matrix **M** has been multiplied by its corresponding eigenvalue. This same effect can be obtained by postmultiplying **M** by a diagonal matrix:

$$\mathbf{AM} = [\mathbf{x}_1\ \mathbf{x}_2 \ldots, \mathbf{x}_n] \begin{bmatrix} \lambda_1 & 0 & 0 \ldots & 0 \\ 0 & \lambda_2 & 0 \ldots & 0 \\ \cdot & \cdot & \cdot \ldots & \cdot \\ \cdot & \cdot & \cdot \ldots & \cdot \\ \cdot & \cdot & \cdot \ldots & \cdot \\ 0 & \cdot & \cdot \ldots & \lambda_n \end{bmatrix} = \mathbf{MS}$$

Where **S** is the spectral matrix.

If we premultiply both sides of the expression by $\mathbf{M}^{-1}$ we obtain

$$\mathbf{M}^{-1}\mathbf{AM} = \mathbf{S}$$

and we see that the matrix **A** can be transformed into a diagonal matrix. The same expression applies for the normalised modal matrix, Hence

$$\mathbf{N}^{-1}\mathbf{AN} = \mathbf{S}$$

It should be noted that for a square matrix of dimensions $n \times n$ to be diagonalised it must have n different eigenvalues or be symmetrical. As this will be the condition of the matrices discussed in this book the matter will not be considered further.

ORTHOGONAL MATRIX

This important matrix is used in both eigenvalue evaluation and coordinate transformations. Briefly an orthogonal matrix is square and its inverse is equal to its transpose, i.e., in an orthogonal matrix **A**, $\mathbf{A}^{-1} = \mathbf{A}^t$.

The arrangements within the matrix that make this condition possible are such that an orthogonal matrix fairly bristles with properties. To begin our examination let us look at a general third order orthogonal matrix and determine what internal properties of the matrix make it possible for its inverse to be equal to its transpose. Let

$$\mathbf{A} = \begin{bmatrix} a_{11} & a_{12} & a_{13} \\ a_{21} & a_{22} & a_{23} \\ a_{31} & a_{32} & a_{33} \end{bmatrix}$$

Then

$$\mathbf{A}^t = \begin{bmatrix} a_{11} & a_{21} & a_{31} \\ a_{12} & a_{22} & a_{32} \\ a_{13} & a_{23} & a_{33} \end{bmatrix} = \mathbf{A}^{-1} \text{ (by definition)}$$

Now

$$\mathbf{AA}^{-1} = \mathbf{I} = \begin{bmatrix} 1 & 0 & 0 \\ 0 & 1 & 0 \\ 0 & 0 & 1 \end{bmatrix}$$

If the formation of the unit matrix is considered it can be seen that the unit terms on the leading diagonal are formed by the sum of the squares of each row of the original matrix **A**.
i.e.

$$a_{11}^2 + a_{12}^2 + a_{13}^2 = 1$$
$$a_{21}^2 + a_{22}^2 + a_{23}^2 = 1$$
$$a_{31}^2 + a_{32}^2 + a_{33}^2 = 1$$

Hence the rows of matrix **A** represent unit vectors. All the zero terms of the unit matrix are created by multiplying one row of matrix **A** by another row.

When the product of two vectors is zero then these vectors are orthogonal. The matrix therefore consists of a set of orthogonal unit vectors. Such a set of orthogonal unit vectors is said to be orthonormal. We can now list the properties that make a matrix orthogonal. Any two different rows when multiplied together give the product = 0. The same row when multiplied by itself gives the product = 1. The same rules must obviously apply to the columns of the matrix:

Any two different columns when multiplied together give the product = 0. The same column when multiplied by itself gives the product = 1.

An interesting, and fairly obvious, property of a symmetrical orthogonal matrix is that it is itself equal to its transpose and therefore its inverse. The reader might like to chose a numerical example to prove the following statements:

If two square matrices **A** and **B** are orthogonal then their product matrix, **C**, is also orthogonal. This rule applies to any number of orthogonal matrices.

DETERMINANT OF AN ORTHOGONAL MATRIX

In the last chapter it was stated that the determinant of a matrix multiplied by the determinant of its inverse is always unity. In symbols $|\mathbf{A}|\,|\mathbf{A}^{-1}| = 1$, but it has also been stated that $|\mathbf{A}| = |\mathbf{A}^t|$. Hence $|\mathbf{A}|$ and $|\mathbf{A}^t|$ in an orthogonal matrix must have the same values, either $+1$ or -1.

COORDINATE TRANSFORMATIONS

Figure 4.1A shows a vector $\overrightarrow{OD}$ referred to a set of mutually orthogonal axes 01, 02 and 03.

Figure 4.1B shows the same vector $\overrightarrow{OD}$ but this time referred to a second set of axes OA, OB and OC. Although still passing through O this second set of axes has been rotated from the original set. Hence Figure 4.1A shows a vector

$$\mathbf{x} = \begin{bmatrix} x_1 \\ x_2 \\ x_3 \end{bmatrix}$$

whereas Figure 4.1B shows a vector

$$\mathbf{y} = \begin{bmatrix} y_1 \\ y_2 \\ y_3 \end{bmatrix}$$

In Chapter 2 mention was made of the direction cosine. A direction cosine is the cosine of the angle between the two sets of reference axes. Either the acute or the obtuse angle can be considered as both angles have the same value of cosine.

Let the cosine of the angle between OA and 01 be a_{11}†, the cosine of the angle between OA and 02 be a_{21} and the cosine of the angle between OA and 03 be a_{31}.

Hence the component y_1, measured along OA, is related to the first set of components thus:

$$y_1 = a_{11}x_1 + a_{21}x_2 + a_{31}x_3$$

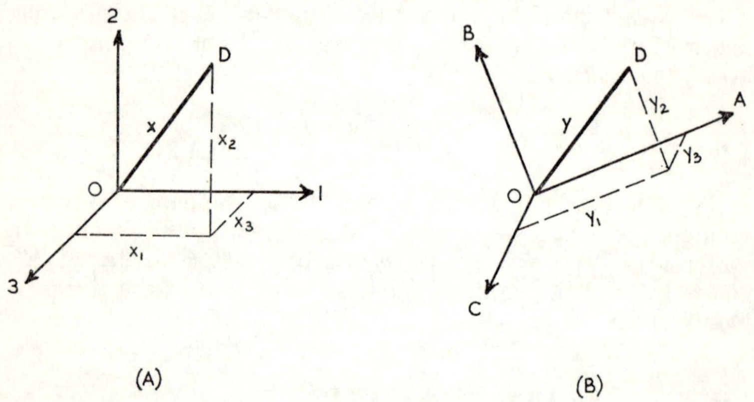

Fig. 4.1.

Similarly the other components y_2 and y_3 are related to the first set of components by:

$$y_2 = a_{12}x_1 + a_{22}x_2 + a_{32}x_3$$

$$y_3 = a_{13}x_1 + a_{23}x_2 + a_{33}x_3$$

These expressions can be written in matrix form:

$$\mathbf{y} = \mathbf{R}^t\mathbf{x}$$

where

$$\mathbf{y} = \begin{bmatrix} y_1 \\ y_2 \\ y_3 \end{bmatrix}, \quad \mathbf{R} = \begin{bmatrix} a_{11} & a_{12} & a_{13} \\ a_{21} & a_{22} & a_{23} \\ a_{31} & a_{32} & a_{33} \end{bmatrix}, \quad \mathbf{x} = \begin{bmatrix} x_1 \\ x_2 \\ x_3 \end{bmatrix}$$

† Note the convention used. The first number of the suffix refers to the first or original set of axes and the second figure refers to the second or rotated set of axes.

R is known as the rotational matrix and consists exclusively of all the relevant direction cosines. Its determinant is always equal to +1.

PROPERTIES OF THE ROTATIONAL MATRIX

As was shown in Chapter 2, in the case of a unit vector, its direction cosines are equal to its components. Hence the sum of the squares of each set of direction cosines in the matrix **R** must equal unity. i.e.

$$a_{11}^2 + a_{21}^2 + a_{31}^2 = 1$$

$$a_{12}^2 + a_{22}^2 + a_{32}^2 = 1$$

$$a_{13}^2 + a_{23}^2 + a_{33}^2 = 1$$

Also if the axes are orthogonal (as they are in Figures 4.1A and B—and generally in stress work) then the sum of the products of the corresponding direction cosines for any two axes must be zero.

If we consider axis OA multiplied by axis OB (i.e. $y_1 y_2$)

$$a_{11}a_{12} + a_{21}a_{22} + a_{31}a_{32} = 0$$

Similarly for OB · OC

$$a_{12}a_{13} + a_{22}a_{23} + a_{32}a_{33} = 0$$

And for OC · OA

$$a_{11}a_{13} + a_{21}a_{23} + a_{31}a_{33} = 0$$

Matrix **R** therefore consists of a set of orthonormal vectors and is therefore an orthogonal matrix.

The discussion so far has concerned the transformation of the components of vector **x** into the components of vector **y** by the relationship $\mathbf{y} = \mathbf{R}^t\mathbf{x}$. Obviously a reverse transformation is possible:

$$\mathbf{x} = [\mathbf{R}^t]^{-1}\mathbf{y} = \mathbf{R}\mathbf{y}$$

($\mathbf{R}^t = \mathbf{R}^{-1}$ in an orthogonal matrix). Hence

$$\mathbf{x} = \begin{bmatrix} a_{11} & a_{12} & a_{13} \\ a_{21} & a_{22} & a_{23} \\ a_{31} & a_{32} & a_{33} \end{bmatrix} \mathbf{y}$$

The above equations can be derived from first principles. The same set of direction cosines still apply as the angles have not changed.

The direction cosines of 01, with respect to OA, OB and OC are a_{11}, a_{12} and a_{13}, etc., and **R** can be obtained.

MATRIX TRANSFORMATIONS

Consider two three-dimensional vectors **x** and **y** related by the expression $\mathbf{y} = \mathbf{Ax}$, where **A** is a square matrix.

The components of **x** (x_1, x_2, x_3) and of **y** (y_1, y_2, y_3) are referred to the same set of orthogonal reference axes.

As explained in the previous section the components of **x** and **y** can be referred to another set of reference axes by the use of the rotational matrix **R**.

Let $\mathbf{x}'$ and $\mathbf{y}'$ be the vectors **x** and **y** referred to the new axes. Then $\mathbf{x}' = \mathbf{R}^t\mathbf{x}$ and $\mathbf{y}' = \mathbf{R}^t\mathbf{y}$. Hence $\mathbf{x} = \mathbf{Rx}'$ and $\mathbf{y} = \mathbf{Ry}'$ (As $\mathbf{R}^t = \mathbf{R}^{-1}$). Now $\mathbf{y}' = \mathbf{R}^t\mathbf{y} = \mathbf{R}^t\mathbf{Ax} = \mathbf{R}^{-1}\mathbf{ARx}'$, i.e. the components of the vectors **x** and **y**, in the new system of coordinates, are related by the expression:

$$\mathbf{y}' = \mathbf{Bx}'$$

where $\mathbf{B} = \mathbf{R}^{-1}\mathbf{AR}$.

The matrices **A** and **B** are said to be congruent and the transformations effected by either of them are known as congruent transformations.

PRINCIPAL AXES TRANSFORMATION

By a suitable choice of the new coordinate system it is possible to arrange for **B** to be the diagonal matrix **S**, i.e.

$$\mathbf{y}' = \mathbf{Bx}' \quad \text{becomes} \quad \mathbf{y}' = \mathbf{Sx}'$$

where **S** = the spectral matrix.

$$\begin{bmatrix} y_1' \\ y_2' \\ \cdot \\ \cdot \\ \cdot \\ y_n' \end{bmatrix} = \begin{bmatrix} \lambda_1 & 0 & 0 & \dots & 0 \\ 0 & \lambda_2 & 0 & \dots & 0 \\ \cdot & \cdot & \cdot & \dots & \cdot \\ \cdot & \cdot & \cdot & \dots & \cdot \\ \cdot & \cdot & \cdot & \dots & \cdot \\ 0 & \cdot & \cdot & \dots & {}_n \end{bmatrix} \begin{bmatrix} x_1' \\ x_2' \\ \cdot \\ \cdot \\ \cdot \\ x_n' \end{bmatrix}$$

The reference axes that form this unique system of coordinates are called the principal axes or latent roots of matrix **A**.

It has been shown that, for this transformation, $\mathbf{N}^{-1}\mathbf{AN} = \mathbf{S}$ where **N** is the modal matrix consisting of normalised eigenvectors.

The elements of matrix **N**, a_{11}, a_{12}, etc. are the direction cosines of the principal axes to the three original reference axes.

Example

In the previous example the modal matrix formed from the normalised eigenvectors, **N**, was found to be:

$$\mathbf{N} = \begin{bmatrix} 0{\cdot}708 & -0{\cdot}658 & 0{\cdot}262 \\ -0{\cdot}603 & -0{\cdot}366 & 0{\cdot}71 \\ -0{\cdot}37 & 0{\cdot}660 & 0{\cdot}657 \end{bmatrix}$$

For the relationship $\mathbf{y} = \mathbf{Ax}$ the expression for a coordinate transformation is $\mathbf{y}' = \mathbf{Bx}'$ and, in the special case of a transformation to the coordinates of the principal axes, is $\mathbf{y}' = \mathbf{Sx}'$. Now $\mathbf{B} = \mathbf{R}^{-1}\mathbf{AR}$ and $\mathbf{S} = \mathbf{N}^{-1}\mathbf{AN}$, hence **N** is equivalent to the rotational matrix **R**, and the directions of the three principal axes are found from the normalised eigenvectors which give the direction cosines. In other words, if

$$\mathbf{N} = \begin{bmatrix} a_{11} & a_{12} & a_{13} \\ a_{21} & a_{22} & a_{23} \\ a_{31} & a_{32} & a_{33} \end{bmatrix}$$

then the elements a_{11}, a_{21} and a_{31} are the direction cosines of the major principal axis to the original reference axes, 1, 2 and 3, etc. Hence the rotation of the principal axes can be obtained directly from matrix **N**. The significance of this result is discussed more fully in chapter 5.

As a point of interest the relationship $\mathbf{N}^{-1}\mathbf{AN} = \mathbf{S}$ can be verified numerically from the example, although this procedure is not necessary for the solution of the problem.

$$\mathbf{AN} = \begin{bmatrix} 12 & -5 & 2 \\ -5 & 9 & -4 \\ 2 & -4 & 7 \end{bmatrix} \begin{bmatrix} 0{\cdot}708 & -0{\cdot}658 & 0{\cdot}262 \\ -0{\cdot}603 & -0{\cdot}366 & 0{\cdot}71 \\ 0{\cdot}37 & 0{\cdot}66 & 0{\cdot}657 \end{bmatrix}$$

$$= \begin{bmatrix} 12{\cdot}251 & -4{\cdot}746 & 0{\cdot}908 \\ -10{\cdot}447 & -2{\cdot}644 & 2{\cdot}452 \\ 6{\cdot}418 & 4{\cdot}768 & 2{\cdot}283 \end{bmatrix}$$

If the reader applies the rule to determine whether or not a matrix is orthogonal he will find that the matrix **N** is indeed orthogonal. Hence

$$\mathbf{N}^{-1} = \mathbf{N}^{t}$$

$$\mathbf{N}^{-1}\mathbf{AN} = \begin{bmatrix} 0{\cdot}708 & -0{\cdot}603 & 0{\cdot}37 \\ -0{\cdot}658 & -0{\cdot}366 & 0{\cdot}660 \\ 0{\cdot}262 & 0{\cdot}71 & 0{\cdot}657 \end{bmatrix} \begin{bmatrix} 12{\cdot}251 & -4{\cdot}746 & 0{\cdot}908 \\ -10{\cdot}447 & -2{\cdot}644 & 2{\cdot}452 \\ 6{\cdot}418 & 4{\cdot}768 & 2{\cdot}283 \end{bmatrix}$$

$$= \begin{bmatrix} 17{\cdot}32 & 0{\cdot}00 & 0{\cdot}00 \\ 0{\cdot}00 & 7{\cdot}22 & 0{\cdot}00 \\ 0{\cdot}00 & 0{\cdot}00 & 3{\cdot}46 \end{bmatrix}$$

5

Forces and Stresses

THERE are two kinds of forces that may act on a body.

SURFACE FORCES

Forces applied from outside the body, e.g. foundation loadings, wind forces, hydrostatic forces, etc.

BODY FORCES

Forces applied within the body, e.g. gravitational forces, seepage forces, etc.

These forces set up stresses within the medium and, if an arbitrary plane within it is considered we can examine the stresses to which it is subjected.

COMPLEX STRESS

If a plane is subjected to a resultant stress that is inclined to the normal to the plane then this stress is referred to as a complex stress.

NORMAL STRESS

The normal stress acting on a plane is simply the component of the complex stress which acts normal to the plane.

SHEAR OR TANGENTIAL STRESS

The shear stress is the component of the complex stress which acts tangentially to the plane. This shear stress can itself be resolved into components. In stress analysis we are dealing with three mutually orthogonal axes and the normal procedure is therefore to resolve the shear stress vector into two components, at right angles to each other, parallel to the two reference axes that the plane lies within.

SYMBOLS AND SIGN CONVENTION

σ—the Greek letter sigma is used to designate both normal and resultant stresses. If there is a risk of ambiguity then a suffix is added. σ_n = normal stress, σ_r = resultant stress.
τ—the Greek letter tau is used to designate shear stress.

If it is necessary to indicate the plane on which these stresses act then suffixes are added.

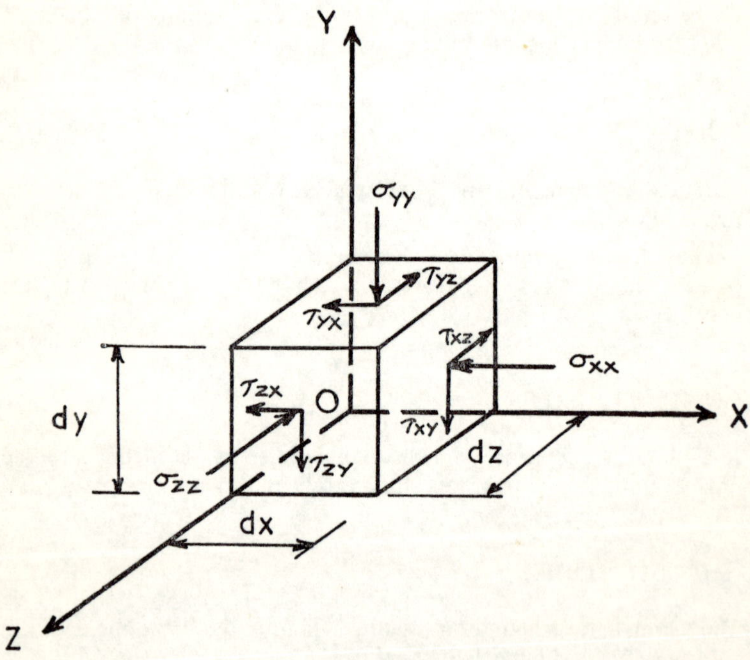

Fig. 5.1.

The usual sign convention used in soil and rock mechanics and the one used in this chapter is:

Compressive stresses—taken as positive.

Tensile stresses—taken as negative.

Shear stresses—when acting on faces furthest from the origin are taken as positive when acting in a direction towards the origin.

Figure 5.1 shows an elemental cube acted upon by the maximum number of stress components possible. In the figure all stresses are positive.

In Figure 5.1 assuming that the element is so small that the body force due to gravitation is negligible and taking moments about the x-axis:

$$\tau_{yz}\,\mathrm{d}x\,\mathrm{d}z\,\mathrm{d}y - \tau_{zy}\,\mathrm{d}x\,\mathrm{d}y\,\mathrm{d}z = 0$$

(Stress) × (area) × (lever arm) = (Stress) × (area) × (lever arm). Hence

$$\tau_{yz} = \tau_{zy}$$

And, similarly:

$$\tau_{xy} = \tau_{yx} \quad \text{and} \quad \tau_{zx} = \tau_{xz}$$

Hence, at any point in a stressed mass we can say that there will be three independent shear stress components and three independent normal stress components:

$$\begin{matrix} \sigma_{xx} & \tau_{xy} \\ \sigma_{yy} & \tau_{xz} \\ \sigma_{zz} & \tau_{yz} \end{matrix}$$

The stresses on the element can be evaluated from these six values which are referred to as the components of stress. Note how the suffixes are applied for the shear stress components. The first suffix is the normal to the plane on which the stress acts, the second suffix is the reference axis in which direction the stress acts.

DETERMINATION OF STRESS VALUES FROM KNOWN STRESS VALUES

Often a system of stresses is either known or can be measured and a common problem is, given this system, to evaluate values of stresses acting on other planes.

PLANE STRESS

Figure 5.1 showed a general three dimensional stress system. In engineering problems it is often found that planes perpendicular to one of the reference axes are not subjected to stress. Such a condition is known as the plane stress condition and we will consider this situation first.

A plane stress system is often referred to as a biaxial stress system. Such a system is shown in Figure 5.2A. The problem is to determine the values of the normal stress, σ, and the shear stress, τ, acting on a plane whose normal, 0A, is inclined at angle θ to the reference axis 01.

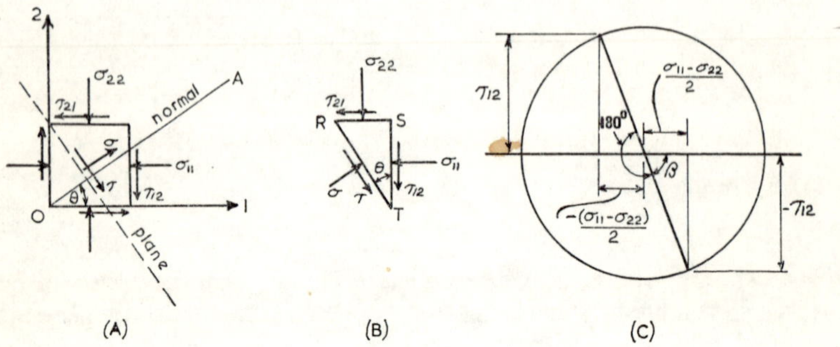

Fig. 5.2.

The system of stresses is shown in Figure 5.2B.

$$RS = RT \sin\theta \quad \text{and} \quad ST = RT\cos\theta$$

If we assume that the width of the triangle normal to the plane of the paper is unity then, by resolving forces parallel to 0A:

$$\sigma RT - \tau_{21}RT\sin\theta\cos\theta - \sigma_{22}RT\sin^2\theta - \tau_{12}RT\cos\theta\sin\theta - \sigma_{11}RT\cos^2\theta = 0$$

i.e.

$$\sigma = \sigma_{11}\cos^2\theta + \sigma_{22}\sin^2\theta + 2\tau_{12}\sin\theta\cos\theta \qquad \text{(1A)}$$

$$\sigma = \frac{\sigma_{11}+\sigma_{22}}{2} + \frac{\sigma_{11}-\sigma_{22}}{2}\cos 2\theta + \tau_{12}\sin 2\theta \qquad (1)$$

Similarly, by resolving forces at right angles to 0A

$$\tau = \frac{\sigma_{11}-\sigma_{22}}{2}\sin 2\theta - \tau_{12}\cos 2\theta \qquad (2)$$

Expression (2) becomes of further interest when we consider what values of θ make τ equal to zero.
For $\tau = 0$:

$$\frac{\sigma_{11}-\sigma_{22}}{2}\sin 2\theta = \tau_{12}\cos 2\theta$$

$$\therefore \quad \tan 2\theta = \frac{\tau_{12}}{(\sigma_{11}-\sigma_{22})/2} = \tan\beta \qquad (3)$$

There will be two values for β differing by 180°. This condition is shown in Figure 5.2C. It will be noticed in the diagram that one of the shear stresses is plotted as negative, although both are positive. This convention is explained later under the heading Mohr's circle of stress.

PRINCIPAL PLANES AND STRESSES

As there are two solutions for β there will be two values of θ:

$$\theta = \frac{\beta}{2} \quad \text{and} \quad \theta = \frac{\beta + 180°}{2} = \frac{\beta}{2} + 90°$$

hence there are two planes, at right angles to each other, on which only normal stresses will act. These planes are known as principal planes and the associated normal stresses are termed principal stresses.

An expression for the magnitude of the principal stresses can be obtained directly from Figure 5.2C. Once the Mohr circle diagram of stress has been studied it will be seen that θ is the angle between 01 and the major principal stress axis and $\theta + 90°$ is the angle between 01 and the minor principal stress axis.

The hypotenuses of the right angled triangles are both equal to:

$$\sqrt{\left\{\left(\frac{\sigma_{11} - \sigma_{22}}{2}\right)^2 + \tau_{12}{}^2\right\}}$$

Hence

$$\sin\beta = \frac{\pm\tau_{12}}{\sqrt{\{[(\sigma_{11} - \sigma_{22})/2]^2 + \tau_{12}{}^2\}}}$$

and

$$\cos\beta = \frac{\pm(\sigma_{11} - \sigma_{22})/2}{\sqrt{\{[(\sigma_{11} - \sigma_{22})/2]^2 + \tau_{12}{}^2\}}}$$

Substituting in Equation (1) the principal stresses are:

$$\sigma = \frac{\sigma_{11} + \sigma_{22}}{2} + \frac{\sigma_{11} - \sigma_{22}}{2}\cos\beta + \tau_{12}\sin\beta$$

$$= \frac{\sigma_{11} + \sigma_{22}}{2} \pm \frac{(\sigma_{11} - \sigma_{22})/2 \cdot (\sigma_{11} - \sigma_{22})/2}{\sqrt{\{[(\sigma_{11} - \sigma_{22})/2]^2 + \tau_{12}{}^2\}}}$$

$$\pm \frac{\tau_{12} \cdot \tau_{12}}{\sqrt{[(\sigma_{11} - \sigma_{22})/2]^2 + \tau_{12}}}$$

i.e.

$$\left.\begin{matrix}\sigma_1\\ \sigma_3\end{matrix}\right\} = \frac{\sigma_{11}+\sigma_{22}}{2} \pm \sqrt{\left\{\left(\frac{\sigma_{11}-\sigma_{22}}{2}\right)^2 + \tau_{12}{}^2\right\}}$$

In a three dimensional stress system there are three principal stresses, the major, the intermediate and the minor. These stresses are designated as σ_1, σ_2 and σ_3 respectively. In a two dimensional system it is the intermediate stress that is omitted and the above formula gives the values of the major principal stress, σ_1 and the minor principal stress, σ_3.

The principal planes are named after the stresses that act upon them. The major principal stress acts on the major principal plane, the intermediate principal stress on the intermediate principal plane, etc.

A reverse set of equations can easily be formulated whereby, given the principal stresses σ_1 and σ_3, it is possible to evaluate σ and τ acting on any plane.

If in Equations (1) and (2) we consider σ_{11} and σ_{22} to be the principal stresses σ_1 and σ_3 then τ_{12} will be zero. 01 will represent the minor principal plane and 02 will represent the major principal plane.

Hence the expressions for σ and τ give the normal and shear stresses acting on a plane whose normal is inclined at angle θ to the major principal stress. (This is the same as saying that the plane is inclined at θ to the major principal plane.) Substituting in (1):

$$\sigma = \frac{\sigma_1+\sigma_3}{2} + \frac{\sigma_1-\sigma_3}{2}\cos 2\theta$$

$$= \sigma_3 + (\sigma_1 - \sigma_3)\cos^2\theta \tag{5}$$

Substituting in (2):

$$\tau = \frac{\sigma_1-\sigma_3}{2}\sin 2\theta \tag{6}$$

DIRECTION COSINES

The direction of a plane is normally specified by the cosines of the angles that the normal to the plane makes with the three reference axes. These values are known as direction cosines and are often given the symbols l, m and n or a_{1r}, a_{2r} and a_{3r}.
In Figure 5.3

$l = a_{1r} = \cos\alpha =$ cosine of angle between 0R and 01

$m = a_{2r} = \cos\beta =$ cosine of angle between 0R and 02

$n = a_{3r} = \cos\gamma =$ cosine of angle between 0R and 03

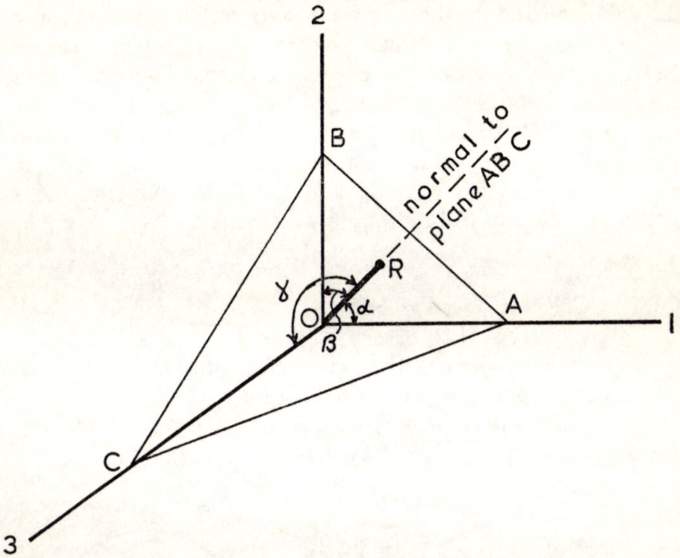

Fig. 5.3.

Either the acute or the obtuse angle can be taken as both angles have the same value of cosine.

It is important not to confuse principal planes with principal stress axes. In Figure 5.4A the major and minor principal planes, $0X$ and $0Y$ are shown.

Now the major principal stress acts at right angles to the major principal plane. Hence $0Y$ gives the direction of both the minor

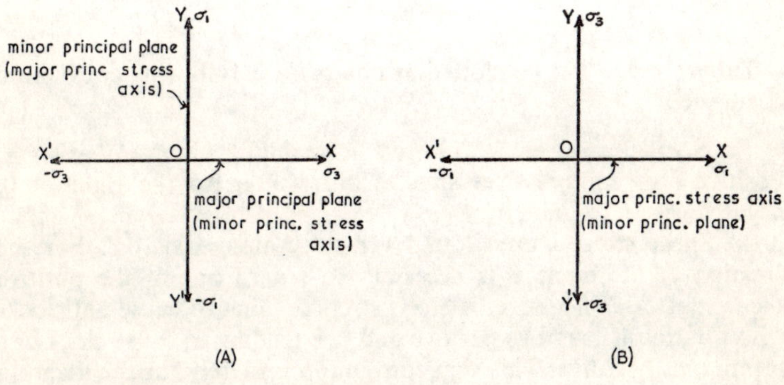

Fig. 5.4.

principal plane and the major principal stress. The convention shown in Figure 5.4A is that generally used in soil-mechanics text books.

In stress analysis work, where we are concerned with axis rotation, the usual convention is to call the major principal stress axis $0X$ and the stress diagram is plotted with the reference axes shown in Figure 5.4B. Although the two conventions lead to the same results the situation can be confusing and should be remembered when reference is made to soil-mechanics text books.

MOHR'S CIRCLE OF STRESS

This is a method whereby the derived Equations (1) to (6) can be expressed in graphical form by means of a circle.

Suppose that an elemental lamina is subjected to the positive stresses σ_{11}, τ_{12} and σ_{22}, τ_{21} as shown in Figure 5.5A.

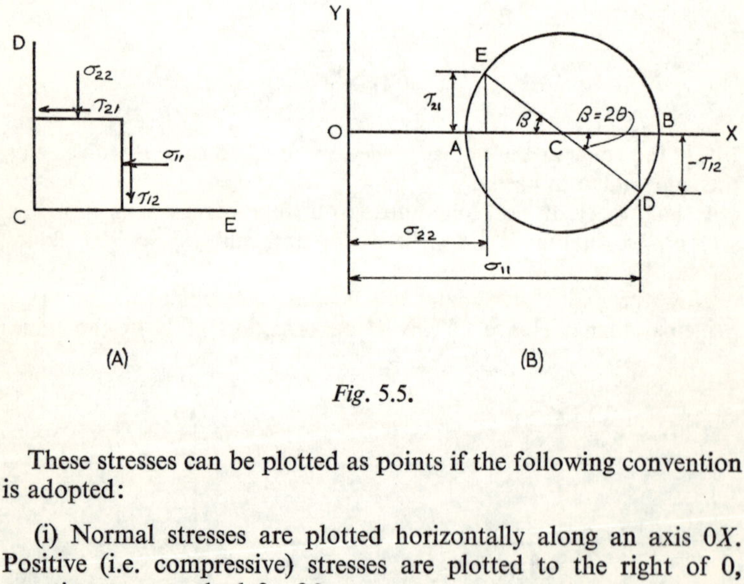

Fig. 5.5.

These stresses can be plotted as points if the following convention is adopted:

(i) Normal stresses are plotted horizontally along an axis $0X$. Positive (i.e. compressive) stresses are plotted to the right of 0, negative stress to the left of 0.

(ii) Shear stresses are plotted vertically along axis $0Y$. For the creation of the circle it is necessary to regard one of the positive shear stresses as negative. A shear stress tending to cause anticlockwise rotation is taken as positive and one tending to cause clockwise rotation as negative. This sign convention is merely for the diagram and is of no other significance.

Having established the two points a circle can be drawn through them. (Figure 5.5B). This circle represents the state of stress on any plane at a point within the stressed mass. Where the circle cuts the $0X$ axis the shear stress has no value and $0A$ and $0B$ therefore represent the minor and principal stresses of the system, σ_3 and σ_1 respectively.
The following points of the diagram should be understood.

(i) The line EC represents the plane acted upon by σ_{22} and τ_{21}, i.e. the plane CE.
(ii) The line CD represents the plane acted upon by σ_{11} and τ_{12}, i.e. the plane CD.
(iii) The line CB represents the plane acted upon by σ_1, i.e. the major principal plane.
(iv) The line AC represents the plane acted upon by σ_3, i.e. the minor principal plane.

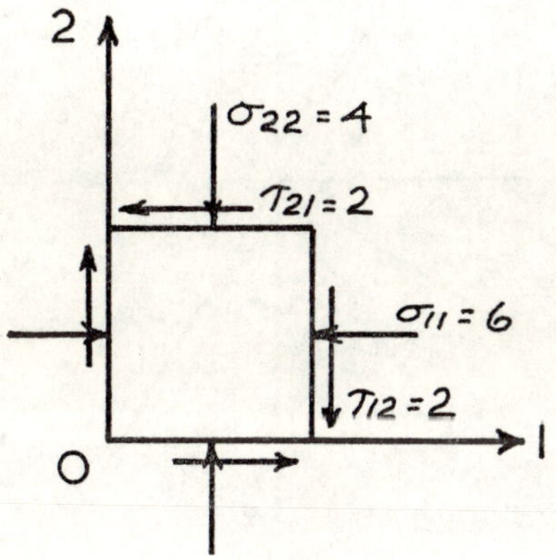

Fig. 5.6.

(v) The planes CE and CD, although physically 90° apart, are represented on the diagram by the lines CE and CD respectively, 180° apart. This is because the angle β in the diagram represents 2θ. The direction of any plane relative to another can be obtained from the diagram by measuring the angle between them and dividing by two. For example the major principal plane is inclined at angle $\beta/2(=\theta)$ anticlockwise from the reference axis CD.

Example

Determine the magnitude and direction of the principal stresses for the biaxial stress system shown in Figure 5.6.

This problem will be solved by three methods.

(1) *By formulae*

$$\left.\begin{matrix}\sigma_1\\ \sigma_3\end{matrix}\right\} = \frac{\sigma_{11}+\sigma_{22}}{2} \pm \sqrt{\left\{\left(\frac{\sigma_{11}-\sigma_{22}}{2}\right)^2 + {\tau_{12}}^2\right\}}$$

$$= \frac{6+4}{2} \pm \sqrt{\left\{\left(\frac{6-4}{2}\right)^2 + 2^2\right\}}$$

$$= 5 \pm \sqrt{(1+4)} = 5 \pm 2{\cdot}236$$

Hence, $\sigma_1 = 7{\cdot}236$; $\sigma_3 = 2{\cdot}764$.

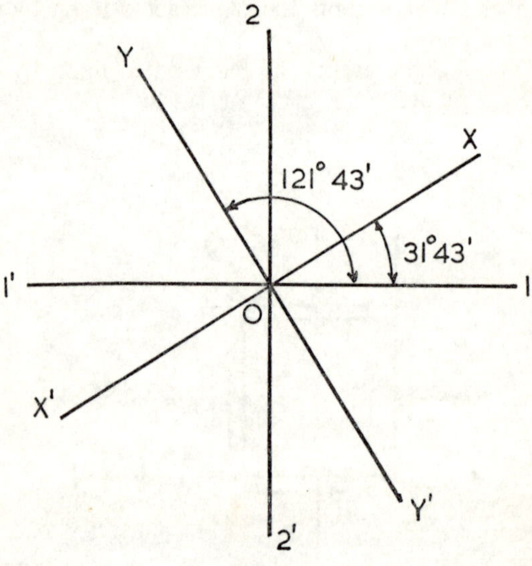

Fig. 5.7.

The directions of the principal stresses can be described in various ways. Adopting the convention of Figure 5.4B, i.e. denoting the major principal stress axis by $0X$ and the minor principal stress axis by $0Y$, the principal stress directions can be specified by the angles between 01 and $0X$, θ, and between 01 and $0Y$, $(\theta + 90°)$.

From Equation (3)

$$\tan\beta = \tan 2\theta = \frac{\tau_{12}}{(\sigma_{11}-\sigma_{22})/2} = \frac{2\times 2}{6-4} = 2$$

$$\therefore \quad 2\theta = 63°26' \quad \text{and} \quad \theta = 31°43'$$

The axis rotation is shown in Figure 5.7.

Hence

$$\sigma_1 = 7{\cdot}236 \text{ inclined at } 31°43' \text{ (anticlockwise) to } 01$$
$$\sigma_3 = 2{\cdot}764 \text{ inclined at } 121°43' \text{ (anticlockwise) to } 01$$

(2) *By matrix algebra*

In formulating Equation (1) we established Equation (1A):

$$\sigma = \sigma_{11}\cos^2\theta + \sigma_{22}\sin^2\theta + 2\tau_{12}\sin\theta\cos\theta$$

This equation can be written in matrix form:

$$\sigma = [\cos\theta \ \sin\theta] \begin{bmatrix} \sigma_{11} & \tau_{12} \\ \tau_{21} & \sigma_{22} \end{bmatrix} \begin{bmatrix} \cos\theta \\ \sin\theta \end{bmatrix}$$

The principal stresses are the roots of the characteristic equation of the stress matrix:

$$\begin{vmatrix} 6-\lambda & 2 \\ 2 & 4-\lambda \end{vmatrix} = 0$$

i.e.

$$\lambda^2 - 10\lambda + 20 = 0$$

The solution of a quadratic of the form $ax^2 + bx + c = 0$ is

$$x = \frac{-b \pm \sqrt{(b^2 - 4ac)}}{2a}$$

Hence

$$\left.\begin{matrix}\sigma_1\\ \sigma_3\end{matrix}\right\} = \frac{10 \pm \sqrt{(100-80)}}{2} = 7{\cdot}236 \text{ and } 2{\cdot}764.$$

Hence

$$\sigma_1 = 7{\cdot}236; \quad \sigma_3 = 2{\cdot}764.$$

To determine the rotation of the axes we must obtain the normalised modal matrix **N**. Each term of this matrix is a direction cosine. Let

$$\mathbf{N} = \begin{bmatrix} a_{11} & a_{12} \\ a_{21} & a_{22} \end{bmatrix}$$

The eigenvector associated with the eigenvalue corresponding to the major principal stress is found from the expression:

$$\begin{bmatrix} 6-\lambda & 2 \\ 2 & 4-\lambda \end{bmatrix} \begin{bmatrix} a_{11} \\ a_{21} \end{bmatrix} = 0$$

i.e.

$$\begin{bmatrix} -1{\cdot}236 & 2 \\ 2 & -3{\cdot}236 \end{bmatrix} \begin{bmatrix} a_{11} \\ a_{21} \end{bmatrix} = 0$$

i.e.

$$-1{\cdot}236a_{11} + 2a_{21} = 0$$
$$2a_{11} - 3{\cdot}236a_{21} = 0$$
$$a_{11} = \frac{2}{1{\cdot}236}a_{21} = 1{\cdot}618a_{21}$$

Putting $a_{21} = 1{\cdot}0$, $a_{11} = 1{\cdot}618$ and the eigenvector is

$$\begin{bmatrix} 1{\cdot}618 \\ 1{\cdot}0 \end{bmatrix}$$

$$\text{Length of vector} = \sqrt{(1{\cdot}618^2 + 1{\cdot}0^2)} = 1{\cdot}902$$

∴ The normalised eigenvector is

$$\begin{bmatrix} 0{\cdot}851 \\ 0{\cdot}525 \end{bmatrix}$$

Similarly for $\lambda = 2{\cdot}764$, (σ_3) the normalised eigenvector is

$$\begin{bmatrix} -0{\cdot}525 \\ 0{\cdot}851 \end{bmatrix}$$

Hence

$$\mathbf{N} = \begin{bmatrix} a_{11} & a_{12} \\ a_{21} & a_{22} \end{bmatrix} = \begin{bmatrix} 0{\cdot}851 & -0{\cdot}525 \\ 0{\cdot}525 & 0{\cdot}851 \end{bmatrix}$$

Now, a_{11} is the direction cosine between $0X$ and 01, a_{21} is the direction cosine between $0X$ and 02, a_{12} is the direction cosine between $0Y$ and 01, and a_{22} is the direction cosine between $0Y$ and 02. Hence

$$\cos^{-1}\theta = 0{\cdot}851 \quad \therefore \theta = 31°43'$$

i.e.

$$\sigma_1 = 7{\cdot}236 \text{ inclined at } 31°43' \text{ (anticlockwise) to 01}$$

and

$$\sigma_3 = 2{\cdot}764 \text{ inclined at } 121°43' \text{ (anticlockwise) to 01}$$

Note. There is little doubt that the solution by formulae presents an easier task. However, in three dimensional work the matrix solution becomes the simplest method. This example has been included here by way of illustration of the method and is not intended to indicate a standard method for solving two dimensional problems.

(3) *By Mohr's circle of stress*

For this example the construction is as follows:
Draw axes $0X$ and $0Y$. σ_{11} and σ_{22} are both positive and are therefore plotted on $0X$ to the right of 0.

For the construction τ_{21} is regarded as positive and plotted above $0X$ whereas τ_{12} is regarded as negative and plotted below $0X$.

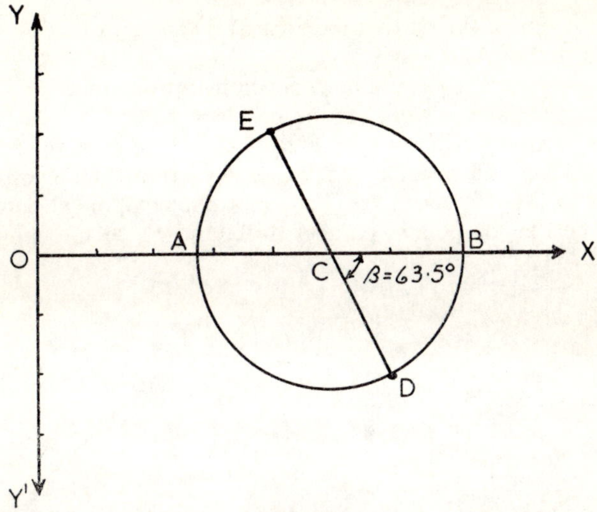

Fig. 5.8.

Plot the point D (6, −2) to represent the stresses on plane 02, and the point E (4, 2) to represent the stresses on plane 01.

Draw the circle through E and D to establish the points A and B on the 0*X* axis (Figure 5.8). Then σ_1 = 0B = 7·25; and σ_3 = 0A = 2·75.

In the diagram CD represents the direction of the plane 02 and CB represents the direction of the major principal plane. The angle between CD and CB = 63·5°.

The major principal plane is at 63·5/2 = 31·75° anticlockwise to 02. ∴ the major principal stress is at (90 + 31·75) = 121·75° anticlockwise to 02, and at (121·75 − 90) = 31·75° anticlockwise to 01.

This will lead to Figure 5.7 if the convention of Figure 5.4B is adopted.

TRIAXIAL STRESS

Figure 5.9 shows an elemental cube subjected to a three dimensional stress system. As proved previously:

$$\tau_{12} = \tau_{21} \qquad \tau_{13} = \tau_{31} \qquad \tau_{23} = \tau_{32}$$

Suppose that the cube is cut by a plane which intersects the cube at the points A, B and C on reference axes 01, 02 and 03 respectively. Let the line 0R, through 0 and normal to the surface ABC have direction cosines: a_{1r} with axis 01, a_{2r} with axis 02 and a_{3r} with axis 03, and let the plane ABC have unit area.

Then it can be shown that the area of B0C = a_{1r}, that of A0C = a_{2r}, and that of A0B = a_{3r}.

The problem is to determine expressions for the normal and shear stress components, σ and τ, acting on plane ABC.

Total force on ABC = $\sigma_r \times$ area = σ_r, where σ_r is the resultant stress on ABC. This force can be resolved into three components parallel to the reference axes. Let F_1 be the component parallel to 01, F_2 be the component parallel to 02 and F_3 be the component parallel to 03.

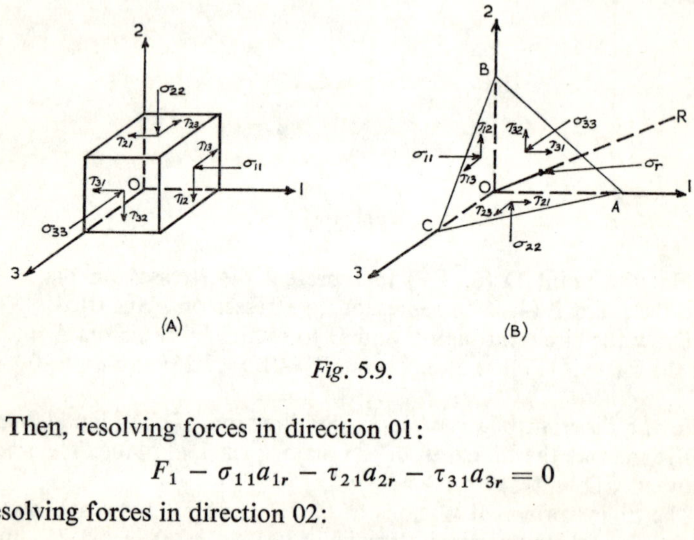

Fig. 5.9.

Then, resolving forces in direction 01:

$$F_1 - \sigma_{11}a_{1r} - \tau_{21}a_{2r} - \tau_{31}a_{3r} = 0$$

resolving forces in direction 02:

$$F_2 - \sigma_{22}a_{2r} - \tau_{32}a_{3r} - \tau_{12}a_{1r} = 0$$

and resolving forces in direction 03:

$$F_3 - \sigma_{33}a_{3r} - \tau_{13}a_{1r} - \tau_{23}a_{2r} = 0$$

The normal component of σ_r is σ and can be found by summing the components of F_1, F_2 and F_3 in direction 0R.

$$\sigma = a_{1r}F_1 + a_{2r}F_2 + a_{3r}F_3$$

$$\begin{aligned}\sigma = a_{1r}(\sigma_{11}a_{1r} + \tau_{21}a_{2r} + \tau_{31}a_{3r}) \\ + a_{2r}(\sigma_{22}a_{2r} + \tau_{32}a_{3r} + \tau_{12}a_{1r}) \\ + a_{3r}(\sigma_{33}a_{3r} + \tau_{13}a_{1r} + \tau_{23}a_{2r}) \\ = \sigma_{11}a_{1r}^2 + \tau_{21}a_{1r}a_{2r} + \tau_{31}a_{1r}a_{3r} \\ + \sigma_{22}a_{2r}^2 + \tau_{32}a_{2r}a_{3r} + \tau_{12}a_{2r}a_{1r} \\ + \sigma_{33}a_{3r}^2 + \tau_{13}a_{3r}a_{1r} + \tau_{23}a_{3r}a_{2r}\end{aligned} \qquad (7)$$

Which may be expressed in the matrix form:

$$\sigma = [a_{1r} \quad a_{2r} \quad a_{3r}] \begin{bmatrix} \sigma_{11} & \tau_{21} & \tau_{31} \\ \tau_{12} & \sigma_{22} & \tau_{32} \\ \tau_{13} & \tau_{23} & \sigma_{33} \end{bmatrix} \begin{bmatrix} a_{1r} \\ a_{2r} \\ a_{3r} \end{bmatrix}$$

Now

$$\tau_{12} = \tau_{21}; \quad \tau_{23} = \tau_{32}; \quad \tau_{13} = \tau_{31}.$$

Hence:

$$\sigma = [a_{1r} \quad a_{2r} \quad a_{3r}] \begin{bmatrix} \sigma_{11} & \tau_{12} & \tau_{13} \\ \tau_{21} & \sigma_{22} & \tau_{23} \\ \tau_{31} & \tau_{32} & \sigma_{33} \end{bmatrix} \begin{bmatrix} a_{1r} \\ a_{2r} \\ a_{3r} \end{bmatrix}$$

The shear stress, τ, acting on ABC is simply the vector difference between the resultant stress, σ_r, and the normal stress component, σ. The resultant stress is the vector sum of F_1, F_2 and F_3 divided by the area of ABC which has been taken as unity.
Hence

$$\tau = \sqrt{(\sigma_r^2 - \sigma^2)}$$

$$\therefore \quad \tau = \sqrt{[F_1^2 + F_2^2 + F_3^2 - (a_{1r}F_1 + a_{2r}F_2 + a_{3r}F_3)^2]} \quad (8)$$

MOHR'S STRESS CIRCLE DIAGRAM FOR TRIAXIAL STRESS

Let σ_1, σ_2 and σ_3 be the principal stresses acting at a point in a medium and let l, m and n be the direction cosines of the normal to a plane through the point to the principal stress axes $0X$, $0Y$ and $0Z$. From Equation (7), remembering that the shear stress components on principal planes are zero, the normal stress on the plane will be:

$$\sigma = \sigma_1 l^2 + \sigma_2 m^2 + \sigma_3 n^2 \quad (9)$$

And, from Equation (8):

$$\tau^2 = \sigma_r^2 - \sigma^2$$
$$= F_1^2 + F_2^2 + F_3^2 - \sigma^2.$$

Now, assuming that the area of the plane is unity, $F_1 = l\sigma_1$, $F_2 = m\sigma_2$ and $F_3 = n\sigma_3$. Hence

$$\tau^2 = l^2\sigma_1^2 + m^2\sigma_2^2 + n^2\sigma_3^2 - \sigma^2 \quad (10)$$

l, m and n are the components of a normalised eigenvector and, as explained in chapter 4, the sum of their squares is unity. Hence we have three equations:

$$\sigma = \sigma_1 l^2 + \sigma_2 m^2 + \sigma_3 n^2$$
$$\sigma^2 + \tau^2 = l^2\sigma_1{}^2 + m^2\sigma_2{}^2 + n^2\sigma_3{}^2$$
$$l = l^2 + m^2 + n^2$$

Solving for l, m and n:

$$l^2 = \frac{(\sigma_2 - \sigma)(\sigma_3 - \sigma) + \tau^2}{(\sigma_2 - \sigma_1)(\sigma_3 - \sigma_1)} \tag{11}$$

$$m^2 = \frac{(\sigma_3 - \sigma)(\sigma_1 - \sigma) + \tau^2}{(\sigma_3 - \sigma_2)(\sigma_1 - \sigma_2)} \tag{12}$$

$$n^2 = \frac{(\sigma_1 - \sigma)(\sigma_2 - \sigma) + \tau^2}{(\sigma_1 - \sigma_3)(\sigma_2 - \sigma_3)} \tag{13}$$

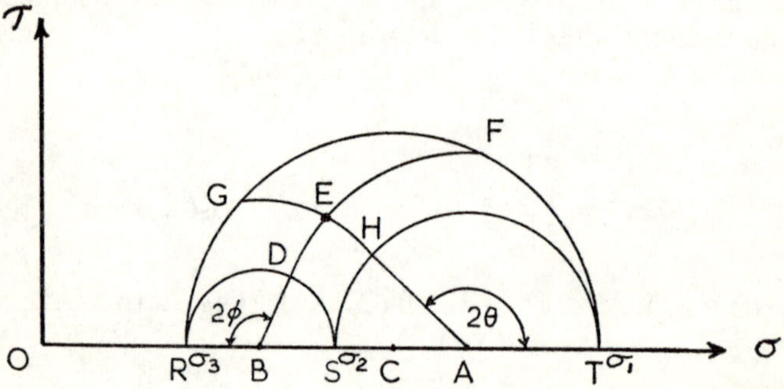

Fig. 5.10.

If we consider all the planes for which n is constant then, from Equation (13) we can evaluate τ in terms of σ. If we plot τ against σ for a fixed value of n we obtain a circle of radius $[\frac{1}{4}(\sigma_1 + \sigma_2)^2 + n^2(\sigma_1 - \sigma_3)(\sigma_2 - \sigma_3)]^{\frac{1}{2}}$ and centre $[\frac{1}{2}(\sigma_1 + \sigma_2), 0]$, i.e., centre A in Figure 5.10.

Now n can have any value from 0 to 1. When $n = 0$ the radius $= (\sigma_1 - \sigma_2)/2$ and when $n = 1$

$$\text{the radius} = \sqrt{[\tfrac{1}{4}(\sigma_1 - \sigma_2)^2 + \sigma_1\sigma_2 - \sigma_1\sigma_3 - \sigma_2\sigma_3 + \sigma_3{}^2]}$$
$$= \sqrt{[\tfrac{1}{4}(\sigma_1 + \sigma_2)^2 - \sigma_1\sigma_3 - \sigma_2\sigma_3 + \sigma_3{}^2]}$$
$$= \tfrac{1}{2}(\sigma_1 + \sigma_2) - \sigma_3.$$

The radius of the circle can therefore vary from AS to AR (in Figure 5.10). Between these limits a variety of circles can be drawn, a typical one, DEF, is shown in Figure 5.10. Similarly, if l is constant, circles can be drawn with centre B and varying in radius from BS to BT. If m is constant, circles of centre C can be drawn with the radii varying from CS to CR.

As $l^2 + m^2 + n^2 = 1$ there is no need to consider more than two sets of circles as the components l and n are sufficient to completely define the system.

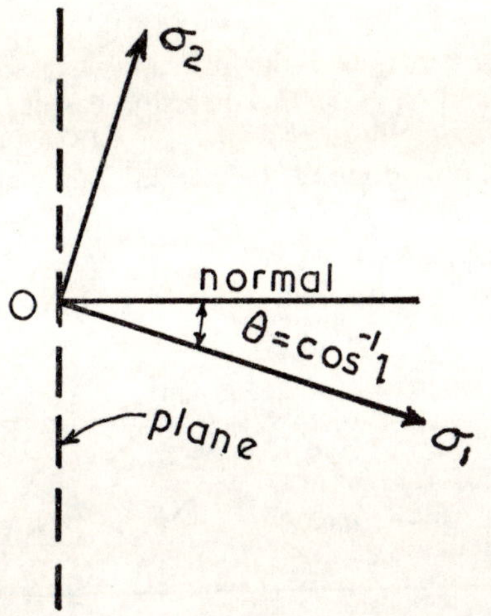

Fig. 5.11.

If we consider these two sets of circles. The circles with centres A and B are the two dimensional stress circles for the (σ_1, σ_2) and (σ_2, σ_3) planes. In the (σ_1, σ_2) circle the point H represents the normal and shear stress components on a plane whose normal is inclined at θ to the major principal stress axis. (Figure 5.11). i.e. $\theta = \cos^{-1} l$. In the (σ_2, σ_3) circle the point D represents the stresses on a plane defined by $\varphi = \cos^{-1} n$.

Obviously where these two circles intersect (the point E) we have coordinates which represent stresses on a plane whose normal has direction cosines n and l. As m is dependent upon the values of n and l the point E represents the stresses on a plane with direction cosines l, m and n.

CONSTRUCTION OF THE MOHR DIAGRAM (given σ_1, σ_2, σ_3; l, m and n)

Plot σ_1, σ_2 and σ_3 on the normal stress axis and draw the three circles. Establish θ from $\cos^{-1} l$ and φ from $\cos^{-1} n$. Draw AH and BD at 2θ and 2φ (as shown in Figure 5.10). With centre A and radius AD draw arc DEF. With centre B and radius BH draw arc HEG.

Now E is the point whose coordinates represent the normal and shear stress components on the plane whose normal has direction cosines to 01, 02 and 03 respectively.

Example

In a triaxial stress system the principal stresses at a point are $\sigma_1 = 8$, $\sigma_2 = 6$ and $\sigma_3 = 3$. The direction cosines of the normal to a plane through the point are $l = 0{\cdot}7$, $m = 0{\cdot}387$ and $n = 0{\cdot}6$.

Determine the magnitude of the normal and shear stresses on the plane.

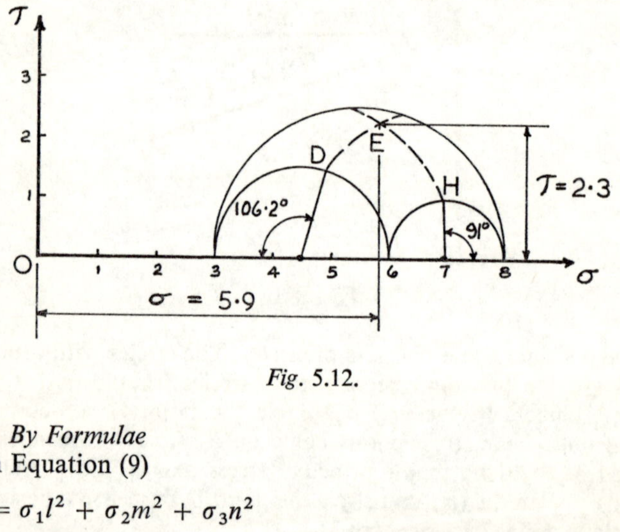

Fig. 5.12.

(1) *By Formulae*

From Equation (9)

$$\sigma = \sigma_1 l^2 + \sigma_2 m^2 + \sigma_3 n^2$$

$$= 8 \times 0{\cdot}7^2 + 6 \times 0{\cdot}387^2 + 3 \times 0{\cdot}6^2 = 5{\cdot}9$$

From Equation (10)

$$\tau = \sqrt{(l^2\sigma_1{}^2 + m^2\sigma_2{}^2 + n^2\sigma_3{}^2 - \sigma^2)}$$

$$= \sqrt{(8^2 + 0{\cdot}7^2 + 6^2 \times 0{\cdot}387^2 + 3^2 \times 0{\cdot}6^2 - 5{\cdot}9^2)} = 2{\cdot}28$$

(2) *By Mohr diagram*

$$\theta = \cos^{-1} l = 45°30' \quad \text{and} \quad \varphi = \cos^{-1} n = 53°8'$$

The construction of the diagram is shown in Figure 5.12. From it $\sigma = 5{\cdot}9$ and $\tau = 2{\cdot}3$.

Example

Referring to Figure 5.9, determine the magnitude and direction of the principal stresses if the applied stress system is:

$$\sigma_{11} = 16 \quad \tau_{12} = -5$$
$$\sigma_{22} = 10 \quad \tau_{23} = 6$$
$$\sigma_{33} = 8 \quad \tau_{13} = -4$$

Now

$$\sigma = [a_{1r} \quad a_{2r} \quad a_{3r}] \begin{bmatrix} \sigma_{11} & \tau_{12} & \tau_{13} \\ \tau_{21} & \sigma_{22} & \tau_{23} \\ \tau_{31} & \tau_{32} & \sigma_{33} \end{bmatrix} \begin{bmatrix} a_{1r} \\ a_{2r} \\ a_{3r} \end{bmatrix}$$

The stress matrix is therefore:

$$\begin{bmatrix} 16 & -5 & -4 \\ -5 & 10 & 6 \\ -4 & 6 & 8 \end{bmatrix} \quad \text{as} \quad \begin{matrix} \tau_{21} = \tau_{12} \\ \tau_{32} = \tau_{23} \\ \tau_{31} = \tau_{13} \end{matrix}$$

The roots of the characteristic equation of the matrix give the principal stresses of the system, i.e.

$$\begin{vmatrix} (16 - \sigma) & -5 & -4 \\ -5 & (10 - \sigma) & 6 \\ -4 & 6 & (8 - \sigma) \end{vmatrix} = 0$$

On expanding this determinant the resulting cubic expression is:

$$\sigma^3 - 34\sigma^2 + 291\sigma - 584 = 0$$

which factorises to

$$(\sigma - 21{\cdot}97)(\sigma - 9{\cdot}11)(\sigma - 2{\cdot}92) = 0$$

Hence

$$\sigma_1 = 21{\cdot}97, \quad \sigma_2 = 9{\cdot}11, \quad \text{and} \quad \sigma_3 = 2{\cdot}92$$

(A method for determining the roots of a cubic expression is given in the appendix, page 202.)

To determine the rotation of the principal stress axes from the reference axes 01, 02 and 03 we must obtain the normalised modal matrix N. Let

$$\mathbf{N} = \begin{bmatrix} a_{11} & a_{12} & a_{13} \\ a_{21} & a_{22} & a_{23} \\ a_{31} & a_{32} & a_{33} \end{bmatrix}$$

To establish the first eigenvector,

$$\begin{bmatrix} a_{11} \\ a_{21} \\ a_{31} \end{bmatrix}$$

we will use the eigenvalue corresponding to the major principal stress. When this operation has been carried out the components of the vector will give the direction of $0X$ relative to 01, 02 and 03. i.e. a_{11} = direction cosine of angle between 01 and $0X$, a_{21} = direction cosine of angle between 02 and $0X$, and a_{31} = direction cosine of angle between 03 and $0X$.

This eigenvector is obtained from the expression:

$$\begin{bmatrix} -5{\cdot}97 & -5 & -4 \\ -5 & -11{\cdot}97 & 6 \\ -4 & 6 & -13{\cdot}97 \end{bmatrix} \begin{bmatrix} a_{11} \\ a_{21} \\ a_{31} \end{bmatrix} = 0$$

$$\therefore \quad -5{\cdot}97a_{11} - 5a_{21} - 4a_{31} = 0 \qquad \text{(A)}$$

$$-5a_{11} - 11{\cdot}97a_{21} + 6a_{31} = 0 \qquad \text{(B)}$$

From (A)

$$a_{11} = \frac{-5}{5{\cdot}97} a_{21} \frac{-4}{5{\cdot}97} a_{31}$$

$$= -0{\cdot}838a_{21} - 0{\cdot}670a_{31}$$

Substituting in (B):

$$5 \times 0{\cdot}838a_{21} + 5 \times 0{\cdot}670a_{31} - 11{\cdot}97a_{21} + 6a_{31} = 0$$

$$\therefore \quad -7{\cdot}77a_{21} + 9{\cdot}35a_{31} = 0$$

Hence

$$a_{21} = 1{\cdot}203a_{31}$$

And

$$a_{11} = -1{\cdot}203 \times 0{\cdot}838a_{31} - 0{\cdot}670a_{31}$$

$$= -1{\cdot}682a_{31}$$

If we put $a_{31} = 1{\cdot}0$ then the eigenvector becomes:

$$\begin{bmatrix} -1{\cdot}682 \\ 1{\cdot}203 \\ 1{\cdot}0 \end{bmatrix}$$

Length of vector $= \surd(1{\cdot}682^2 + 1{\cdot}203^2 + 1{\cdot}0^2) = 2{\cdot}293$.
The normalised vector is

$$\begin{bmatrix} -0{\cdot}733 \\ 0{\cdot}523 \\ 0{\cdot}436 \end{bmatrix}$$

A check on these values can be obtained from the fact that

$$a_{11}{}^2 + a_{21}{}^2 + a_{31}{}^2 = 1$$

$$0{\cdot}733^2 + 0{\cdot}523^2 + 0{\cdot}436^2 = 0{\cdot}537 + 0{\cdot}273 + 0{\cdot}190 = 1{\cdot}000.$$

The second eigenvector is established by inserting $\sigma = \sigma_2 = 9{\cdot}11$ in the equation and the third eigenvector by inserting $\sigma_3 = 2{\cdot}92$.

The normalised modal matrix can be shown to be:

$$\mathbf{N} = \begin{bmatrix} -0{\cdot}733 & 0{\cdot}681 & -0{\cdot}020 \\ 0{\cdot}523 & 0{\cdot}550 & -0{\cdot}653 \\ 0{\cdot}436 & 0{\cdot}487 & 0{\cdot}753 \end{bmatrix}$$

The first column gives the direction cosines of $0X$ to 01, 02 and 03, the second column gives the direction cosines of $0Y$ to 01, 02 and 03, and the third column gives the direction cosines of $0Z$ to 01, 02 and 03.

The problem has now been solved but for interest a check will be made on the solutions by working backwards from the principal stress values now established to the original stress components given, by means of the Mohr diagram. Consider the stress axis 01:

$$l = \cos^{-1}\theta = -0{\cdot}733$$

$$n = \cos^{-1}\varphi = -0{\cdot}02$$

Negative cosine values always seem to cause trouble. There is little problem if it is remembered that if $\cos\theta$ is negative the angle must lie in either the second or the third quadrant, and will have the value 180° plus or minus the angle which has the same positive value of cosine.

In the example

$$\theta = 180° - 42°50' \quad \text{or} \quad 180° + 42°50'$$

$$= 137°10' \quad \text{or} \quad 222°50'$$

$$\therefore \quad 2\theta = 274°20' \quad \text{or} \quad 445°40' \; (=85°40')$$

In order to plot the line AH (in Figure 5.10) the angle 2θ must lie between 0° and 180°. Hence the value of 2θ that must be taken here is 85°40′. Hence, for this construction, simply take twice the angle with cosine 0·733. Similarly $2\varphi = 177°30'$.

The Mohr diagram is shown in Figure 5.13. From it $\sigma_{11} = 16$; $\tau = 6{\cdot}4$. The shear stress component is of course the vector sum of τ_{12} and τ_{13} i.e. $\sqrt{(5^2 + 4^2)} = 6{\cdot}4$. Consider the stress axis 02

$$l = 0{\cdot}523 \qquad 2\theta = 116°56'$$

$$n = -0{\cdot}653 \quad 2\varphi = 98°28'$$

From the diagram $\sigma_{22} = 10$, $\tau = 7{\cdot}8$. Consider stress axis 03

$$l = 0{\cdot}436 \quad 2\theta = 128°18'$$

$$n = 0{\cdot}753 \quad 2\varphi = 82°18'$$

From the diagram $\sigma_{33} = 8$; $\tau = 7{\cdot}2$.

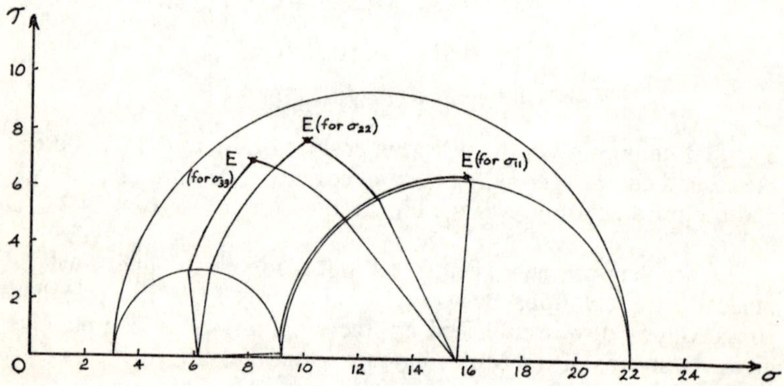

Fig. 5.13.

STRESS INVARIANTS

The characteristic equation of a general three dimensional stress system has been shown to be:

$$\begin{vmatrix} (\sigma_{11} - \sigma) & \tau_{12} & \tau_{13} \\ \tau_{21} & (\sigma_{22} - \sigma) & \tau_{23} \\ \tau_{22} & \tau_{23} & (\sigma_{33} - \sigma) \end{vmatrix} = 0$$

where the reference axes are 01, 02 and 03 and where σ gives the three roots σ_1, σ_2 and σ_3.

Similarly, if the reference axes were OX, OY and OZ the characteristic equation is:

$$\begin{vmatrix} (\sigma_{xx} - \sigma) & \tau_{xy} & \tau_{xz} \\ \tau_{yx} & (\sigma_{yy} - \sigma) & \tau_{yz} \\ \tau_{zx} & \tau_{yz} & (\sigma_{zz} - \sigma) \end{vmatrix} = 0$$

Expanding either of these expressions will give a similar cubic equation. If we expand the first expression the result is:

$$\sigma^3 - (\sigma_{11} + \sigma_{22} + \sigma_{33})\sigma^2 + (\sigma_{11}\sigma_{22} + \sigma_{22}\sigma_{33} + \sigma_{11}\sigma_{33} - \tau_{12}^2 - \tau_{23}^2 - \tau_{31}^2)\sigma$$
$$- (\sigma_{11}\sigma_{22}\sigma_{33} + 2\tau_{12}\tau_{23}\tau_{31} - \sigma_{11}\tau_{23}^2 - \sigma_{22}\tau_{13}^2 - \sigma_3\tau_{12}^2) = 0$$

i.e. $$\sigma^3 - J_1\sigma^2 + J_2\sigma - J_3 = 0$$

Where

$$J = \sigma_{11} + \sigma_{22} + \sigma_{33}$$
$$J_2 = \sigma_{11}\sigma_{22} + \sigma_{22}\sigma_{33} + \sigma_{11}\sigma_{33} - \tau_{12}^2 - \tau_{23}^2 - \tau_{31}^2$$
$$J_3 = \sigma_{11}\sigma_{22}\sigma_{33} + 2\tau_{12}\tau_{23}\tau_{31} - \sigma_{11}\tau_{23}^2 - \sigma_{22}\tau_{13}^2 - \sigma_{33}\tau_{12}^2$$

Now, as the values of σ_{11}, σ_{22} and σ_{33} are constant (provided that the applied stress system is not altered) the cubic expression obtained when the determinant is expanded must also be constant, no matter what set of reference axes and stress components was originally used. Hence J_1, J_2 and J_3 must be constants. They are referred to as stress invariants and given the names first, second and third stress invariants. The values of J_1, J_2 and J_3 can be quickly evaluated from the principal stress values. If the original reference axes had been the principal stress axes then $\sigma_{11} = \sigma_1$, $\sigma_{22} = \sigma_2$ and $\sigma_{33} = \sigma_3$ and there would be no shear stress components. Hence, as J_1, J_2 and J_3 cannot change in value:

$$J_1 = \sigma_1 + \sigma_2 + \sigma_3$$
$$J_2 = \sigma_1\sigma_2 + \sigma_2\sigma_3 + \sigma_3\sigma_1$$
$$J_3 = \sigma_1\sigma_2\sigma_3$$

Example

By means of the stress invariants determine the expansion of the characteristic equation of the stress matrix for Example 5. In the example $\sigma_1 = 21{\cdot}97$, $\sigma_2 = 9{\cdot}11$, $\sigma_3 = 2{\cdot}92$.

$$J_1 = 21{\cdot}97 + 9{\cdot}11 + 2{\cdot}92 = 34{\cdot}0$$
$$J_2 = 21{\cdot}97 \times 9{\cdot}11 + 9{\cdot}11 \times 2{\cdot}92 + 2{\cdot}92 \times 21{\cdot}97 = 291$$
$$J_3 = 21{\cdot}97 \times 9{\cdot}11 \times 2{\cdot}92 = 584$$

Hence

$$\sigma^3 - 34\sigma^2 + 291\sigma - 584 = 0$$

SPACE DIAGONAL AND OCTAHEDRAL PLANE

When considering stresses in a three dimensional system it is general practice to use the stress space formed by the three principal stress axes. In Figure 5.14A the point P gives the values of the three principal stresses.

If we consider all the points at which $\sigma_1 = \sigma_2 = \sigma_3$ then it will be seen that these points will lie on the same straight line going through the origin and making the same angle with each axis ($\cos^{-1} 1/\sqrt{3}$). The condition $\sigma_1 = \sigma_2 = \sigma_3$ is a hydrostatic state of stress and the line OP′ (in Figure 5.14B) is called the hydrostatic stress axis or the space diagonal.

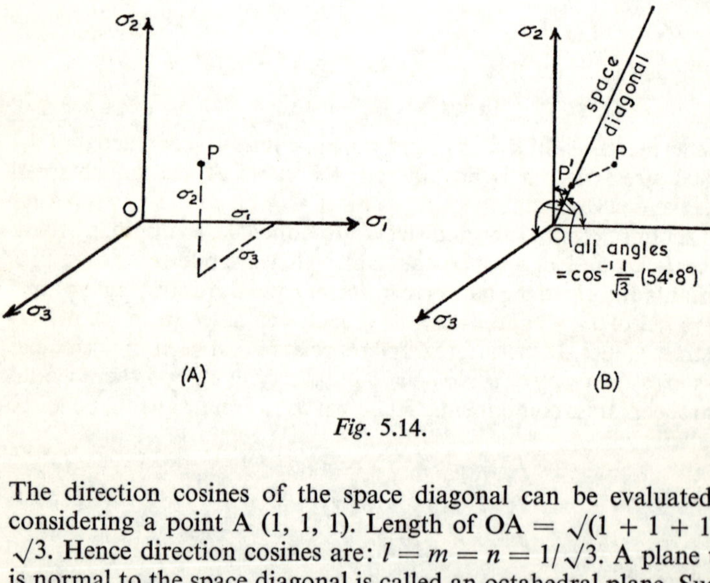

Fig. 5.14.

The direction cosines of the space diagonal can be evaluated by considering a point A (1, 1, 1). Length of OA $= \sqrt{(1 + 1 + 1)} = \sqrt{3}$. Hence direction cosines are: $l = m = n = 1/\sqrt{3}$. A plane that is normal to the space diagonal is called an octahedral plane. Such a plane is illustrated in Figure 5.15. The space diagonal 0R cuts the plane at R. The equation of an octahedral plane can soon be established:

If an arbitrary point, P, of coordinates ($\sigma_1, \sigma_2, \sigma_3$) is chosen on the plane then the component of 0P along the σ_1 axis $=$ 0K $= \sigma_1$, the component of 0P along the σ_2 axis $=$ PM $= \sigma_2$ and the component of 0P along the σ_3 axis $=$ 0L $= \sigma_3$.

Now the projections of these components in the direction of 0R must sum up to the length of 0R. Let the length of $0R = p$. Then

$$0K \cdot l + PM \cdot m + 0L \cdot n = p$$

i.e.

$$\frac{\sigma_1}{\sqrt{3}} + \frac{\sigma_2}{\sqrt{3}} + \frac{\sigma_3}{\sqrt{3}} = p$$

The equation of an octahedral plane is therefore of the form:

$$\sigma_1 + \sigma_2 + \sigma_3 = c$$

The normal stress acting on an octahedral plane is known as the octahedral normal stress and is given the symbol σ_{oct}.

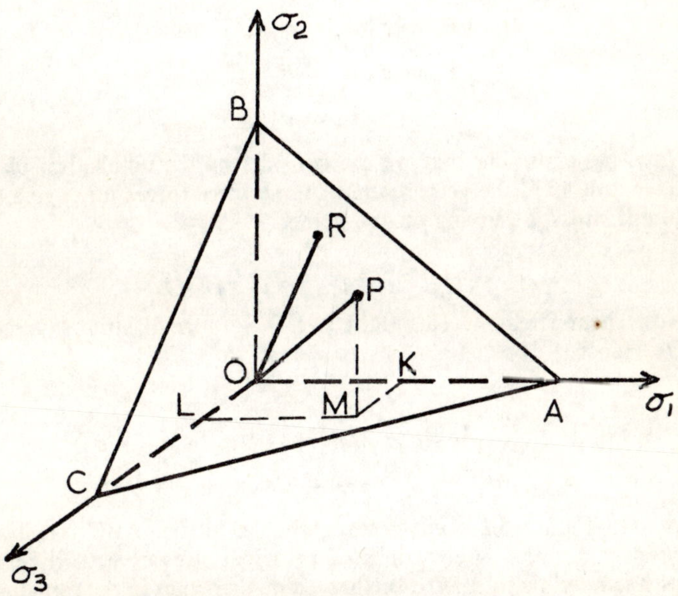

Fig. 5.15.

We have already seen that the normal stress on a plane whose normal has direction cosines a_{1r}, a_{2r} and a_{3r} to the reference axes 01, 02 and 03 can be obtained from the expression:

$$\sigma = (a_{1r} \quad a_{2r} \quad a_{3r}) \begin{bmatrix} \sigma_{11} & \tau_{12} & \tau_{13} \\ \tau_{21} & \sigma_{22} & \tau_{23} \\ \tau_{31} & \tau_{32} & \sigma_{33} \end{bmatrix} \begin{bmatrix} a_{1r} \\ a_{2r} \\ a_{3r} \end{bmatrix}$$

When the reference axes are the principal stress axes $\sigma_{11} = \sigma_1$, $\sigma_{22} = \sigma_2$, $\sigma_{33} = \sigma_3$ and there are no shear stresses.

For an octahedral plane $a_{1r} = a_{2r} = a_{3r} = 1/\sqrt{3}$.

Hence

$$\sigma_{oct} = \left(\frac{1}{\sqrt{3}} \; \frac{1}{\sqrt{3}} \; \frac{1}{\sqrt{3}}\right) \begin{bmatrix} \sigma_1 & 0 & 0 \\ 0 & \sigma_2 & 0 \\ 0 & 0 & \sigma_3 \end{bmatrix} \begin{bmatrix} 1/\sqrt{3} \\ 1/\sqrt{3} \\ 1/\sqrt{3} \end{bmatrix}$$

$$= \tfrac{1}{3}(\sigma_1 + \sigma_2 + \sigma_3)$$

Let F_1, F_2 and F_3 be the forces due to, and acting in the same directions as, σ_1, σ_2 and σ_3.

If we consider the octahedral plane to have unit area then

$$F_1 = \sigma_1 a_{1r} = \sigma_1/\sqrt{3}$$

$$F_2 = \sigma_2 a_{2r} = \sigma_2/\sqrt{3}$$

$$F_3 = \sigma_3 a_{3r} = \sigma_3/\sqrt{3}$$

Now, remembering that we are considering the octahedral plane to have unit area, the vector sum of these three forces must give the resultant stress, σ_r, acting on the plane.

i.e.

$$\sigma_r = 1/\sqrt{3}\,\sqrt{(\sigma_1{}^2 + \sigma_2{}^2 + \sigma_3{}^2)}$$

Now shear stress = (Resultant stress) – (normal stress) (vectorially). Hence

$$\tau_{oct} = \sqrt{(\sigma_r{}^2 - \sigma_{oct}{}^2)}$$

$$= \sqrt{(\tfrac{1}{3}(\sigma_1{}^2 + \sigma_2{}^2 + \sigma_3{}^2) - \tfrac{1}{9}(\sigma_1 + \sigma_2 + \sigma_3)^2]}$$

$$= \tfrac{1}{3}\sqrt{[(\sigma_1 - \sigma_2)^2 + (\sigma_2 - \sigma_3)^2 + (\sigma_3 - \sigma_1)^2]}$$

Note. In Figure 5.15 only one octahedral plane, ABC is shown. Obviously, for the same values of principal stresses with different signs it is possible to draw a further 7 similar planes in the remaining stress spaces. The plane ABC is therefore one triangular face of an eight sided figure (a regular octahedron). Hence the term octahedral.

RELATIONSHIP BETWEEN MAXIMUM SHEARING STRESS AND OCTAHEDRAL SHEAR STRESS

In Figure 5.10 it is seen that the maximum shear stress possible in a three dimensional stress system is $(\sigma_1 - \sigma_3)/2$. This is of course if the convention $\sigma_1 > \sigma_2 > \sigma_3$ is adopted, as in this book. The octahedral shear stress is very close to this value varying from 0·817 to 0·943 times the maximum shear stress. When $\sigma_2 = \sigma_1$ or when $\sigma_2 = \sigma_3$ the higher value applies and when σ_2 is the average of the σ_1 and σ_3 values the lower figure is obtained.

Example

Determine the ratio of the octahedral shear stress to the maximum shear stress for the following three systems:

	σ_1	σ_2	σ_3
(a)	8	8	4
(b)	8	4	4
(c)	8	6	4

(a) $\tau_{max} = \dfrac{8-4}{2} = 2$

$$\tau_{oct} = \tfrac{1}{3}\sqrt{[(8-4)^2 + (4-8)^2]} = 1{\cdot}866$$

$$\frac{\tau_{oct}}{\tau_{max}} = \frac{1{\cdot}866}{2} = 0{\cdot}943.$$

(b) $\tau_{max} = 2$

$$\tau_{oct} = \tfrac{1}{3}\sqrt{[(8-4)^2 + (4-8)^2]} = 1{\cdot}866$$

$$\frac{\tau_{oct}}{\tau_{max}} = 0{\cdot}943.$$

(c) $\tau_{max} = 2$

$$\tau_{oct} = \tfrac{1}{3}\sqrt{[(6-6)^2 + (6-4)^2 + (4-8)^2]} = 1{\cdot}633$$

$$\frac{\tau_{oct}}{\tau_{max}} = \frac{1{\cdot}633}{2} = 0{\cdot}817.$$

It is often convenient to divide a general stress state into two components:

(a) The hydrostatic component, $(\sigma_1 + \sigma_2 + \sigma_3)/3$.
(b) A deviator stress component, accounting for the remainder.

In Figure 5.14B the general state of stress is represented by the point P. The point P′ represents the hydrostatic component and the line PP′ gives the deviator stress tensor.

6

Elastic and Plastic Theories for Metals and Soils

STRESS–STRAIN RELATIONSHIPS OF METALS

ALTHOUGH equally applicable in compression the stress–strain relationships of a metal are usually obtained by subjecting a prepared sample to a tensile test. The sample is made to fail in tension under an increasing load and measurements of extension and corresponding load are taken at set intervals up to failure. A typical nominal stress–strain curve obtained from such a test is shown in Figure 6.1A. Nominal stresses are obtained by dividing the actual load by the original area of the test sample, i.e. there is no allowance for reduction in area due to extension.

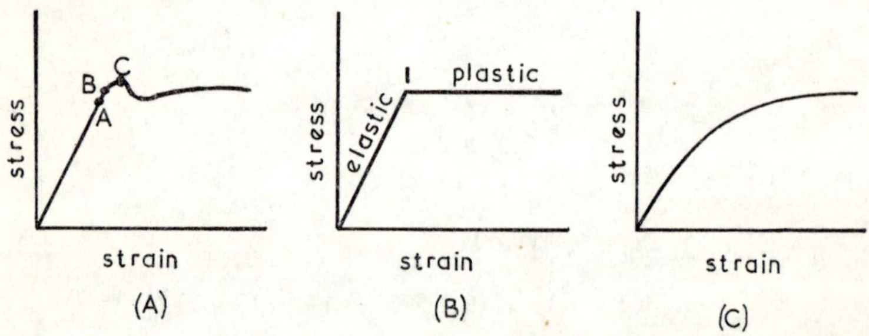

Fig. 6.1.

From the curve it is seen that in the early stages of loading the stress is proportional to the strain. Unloading tests can also demonstrate that in the early loading stages the sample will return to its original dimensions if the load is removed, i.e. it is elastic.

The limiting stress at which elasticity ceases is known as the Elastic Limit, represented by point A in Figure 6.1A.

The limiting stress at which the linearity between stress and strain ceases is known as the limit of proportionality, represented by point B in the figure. In most metals the points A and B occur very close together and are generally regarded as the same point.

The point C in the figure represents the yield point, i.e. the stress at which there is a sudden drop of load, as illustrated, or the stress at which there is a continuing extension under, more or less, the same load. Proof stress is used as the value of the yield stress when there is no definite yield point and is simply the intensity of stress that will cause a certain value of permanent strain, this value is fixed by the designer.

The relationships of Figure 6.1A can be approximated to those of an ideal elastic–plastic material, as shown in Figure 6.1B in which the point 1 represents both the limit of proportionality and the yield stress.

STRESS–STRAIN RELATIONSHIPS OF ROCKS AND SOILS

Rocks and concrete have a similar stress–strain relationship to that represented in Figure 6.1A although the stress in this case can only be compressive. A typical compressive stress–strain curve for a soil is shown in Figure 6.1C. At no point is it really possible to say that there is a linear relationship between stress and strain. One is therefore forced to assume the theoretically idealised conditions shown in Figure 6.1B or to solve a particular problem from test results. In most soil problems the induced stresses are either low enough to assume elastic behaviour (soil settlement) or they are high enough to assume plastic behaviour (bearing capacity and earth-pressure problems).

With soils even further assumptions must be made if one is to obtain a solution. Generally it is assumed that the soil is homogeneous and isotropic. As with the assumption of perfect elasticity these theoretical relationships do not apply in practice but can lead to reasonable results when sensibly applied. The finite element method of stress analysis is a means of solution in which these assumptions can be minimised.

ELASTIC AND PLASTIC THEORIES

As shown in Appendix III the two-dimensional state of stress in an elastic body when weight is the only significant body force can be represented by the differential equation:

$$\frac{\partial^4 \varphi}{\partial x^4} + 2\frac{\partial^4 \varphi}{\partial x^2\,\partial z^2} + \frac{\partial^4 \varphi}{\partial z^4} = 0$$

where φ is the Airy stress function.

Any solution obtained from this equation is valid provided that the stressed material deforms elastically (i.e. linearly). If, at some point in the body, the yield-stress value is exceeded then there are three different zones of stress in evidence:

(1) An elastic or linear stress–strain relationship.

(2) A plastic–elastic stress–strain relationship.

(3) A plastic stress–strain relationship.

If workable solutions are to be obtained we must assume that zone (2) does not exist.

For the elastic zone stress values can be calculated by assuming conditions of elastic equilibrium, for the plastic zone stress values can be calculated by assuming conditions of plastic equilibrium.

ELASTIC BEHAVIOUR

As can be seen from Figure 6.1B there is a linear relationship between stress and strain in the ideal elastic material. By way of revision it will be briefly stated that the component of strain of a body in a particular direction is defined as the deformation per unit original length in the direction considered. i.e. If a bar of original length x is subjected to a tensile force such that the final length of the bar is $x + \delta x$ then the component of strain, in the direction of the bar axis, is $\delta x/x$. In the elastic theory the deformations of a body, extension, shortening, or rotation, are taken to be small and are assumed to vary continuously throughout the volume of the body. These displacements, as with stresses, are generally resolved into three components: u, v and w in the directions parallel to the reference axes 01, 02 and 03 (sometimes referred to as $0X$, $0Y$, $0Z$). The symbol for strain is the Greek letter epsilon, ε, and a strain is described as a tensile, a compressive, or a shear strain, depending upon the stress that causes it.

Shear strain is often given the symbol γ, being the Greek letter gamma. If 01 and 02 are two perpendicular planes through a point 0 in an unstressed mass and these planes change directions to $01'$ and $02'$ in the stressed state so that the angle between them is $(90° - \psi)$ then the shear strain at 0, caused by the stress system is defined as:

$$\gamma = \tan \psi = \psi$$

for small rotations measured in radians.

COMPONENTS OF STRAIN

Consider the elemental cube shown in Fig. 6.2A. Let us assume that the body undergoes a deformation so that u, v and w are the components of displacement of the origin, O.

Consider the displacement of the points O, A and B in the x, y plane. The displacement of A in the x direction will be equal to the displacement at O plus any displacement within the body itself, i.e. $u + (\partial u/\partial x)\,dx$, where $\partial u/\partial x$ is the rate of change of deformation within the body in the direction x. Hence the point A″ is established (Figure 6.2B). Similarly the displacement of B in direction y will be $v + (\partial v/\partial y)\,dy$ to give the point B″. Hence the elongation of OA in

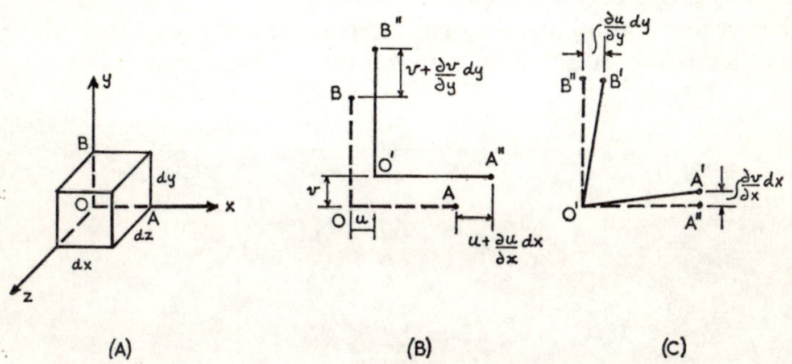

Fig. 6.2.

the x direction is $(\partial u/\partial x)\,dx$ and therefore the strain component in the x direction is:

$$\varepsilon_x = \frac{\partial u}{\partial x}$$

Similarly,

$$\varepsilon_y = \frac{\partial v}{\partial y}$$

and

$$\varepsilon_z = \frac{\partial w}{\partial z}$$

If any part of the displacement vector v becomes effective as movement is made in direction x then the point AA″ must suffer a further displacement, $(\partial v/\partial x)\,dx$, to give the point A′.

I.e. Displacement of A is

$$u + \left(\frac{\partial u}{\partial x}\right) dx + \left(\frac{\partial v}{\partial x}\right) dx$$

Displacement of B is

$$\mathrm{v} + \left(\frac{\partial \mathrm{v}}{\partial y}\right) \mathrm{d}y + \left(\frac{\partial u}{\partial y}\right) \mathrm{d}y$$

The final positions of O, A and B (O′, A′ and B′) are shown in Figure 6.2C. Now, from Figure 6.2C it is obvious that angle BOA, originally a right angle, has decreased by the sum of the angle A′O′A″ and B′O′B″.

Arc length/radius = angle subtended and, ignoring the term $(\partial u/\partial x)\,\mathrm{d}x$:

$$\text{Angle A}'\text{O}'\text{A}'' = \left(\frac{\partial \mathrm{v}}{\partial x}\right)\left(\frac{\mathrm{d}x}{\mathrm{d}x}\right) = \frac{\partial \mathrm{v}}{\partial x}$$

Similarly angle B′O′B″ = $\partial u/\partial y$.

Hence the total angular distortion in plane $(x, y) = (\partial \mathrm{v}/\partial x) + (\partial u/\partial y)$. This then is the shear strain between the (x, z) and the (y, z) planes, i.e.

$$\gamma_{xy} = \frac{\partial \mathrm{v}}{\partial x} + \frac{\partial u}{\partial y}$$

Similarly,

$$\gamma_{xz} = \frac{\partial u}{\partial z} + \frac{\partial w}{\partial x}$$

$$\gamma_{yz} = \frac{\partial w}{\partial y} + \frac{\partial \mathrm{v}}{\partial z}$$

Hence we have obtained:

Tensile (or compressive) strains	*Shear strains*
$\varepsilon_x = \dfrac{\partial u}{\partial x}$	$\gamma_{xy} = \dfrac{\partial \mathrm{v}}{\partial x} + \dfrac{\partial u}{\partial y}$
$\varepsilon_y = \dfrac{\partial \mathrm{v}}{\partial y}$	$\gamma_{xz} = \dfrac{\partial u}{\partial z} + \dfrac{\partial w}{\partial x}$
$\varepsilon_z = \dfrac{\partial w}{\partial z}$	$\gamma_{yz} = \dfrac{\partial w}{\partial y} + \dfrac{\partial \mathrm{v}}{\partial z}$

These six quantities are referred to as components of strain.

RELATIONSHIP BETWEEN COMPRESSIVE/TENSILE STRESSES AND STRAINS

Within the range of linearity the ratio of stress to strain is of constant value and is given the symbol E. E is known as the Modulus of

Elasticity and is defined as:

$$E = \frac{\sigma}{\varepsilon}$$

Consider a rectangular solid with sides parallel to the coordinate axes and subjected to a normal compressive stress, say σ_x, uniformly distributed over two opposite faces.

It has been shown experimentally, in the case of an isotropic material that normal stresses do not produce angular distortion. The magnitude of the strain in the x direction is:

$$\varepsilon_x = \frac{\sigma_x}{E} \qquad (1)$$

This compressive deformation is accompanied by a lateral expansion parallel to the y and z axes:

$$\varepsilon_y = -v\frac{\sigma_x}{E}; \quad \varepsilon_z = -v\frac{\sigma_x}{E} \qquad (2)$$

where v = Poisson's ratio (lateral strain/longitudinal strain). For steel $v \approx 0{\cdot}3$ but for soil its value varies, being taken as 0·5 for a fully saturated soil. These two equations will apply equally in tension or compression as, in an ideal elastic material, E and v have the same values in tension and compression.

If our rectangular element is subjected to the action of three normal stresses, σ_x, σ_y and σ_z, uniformly distributed over the sides, the resultant components of strain can be obtained from Equations (1) and (2) by the principle of superposition:

$$\varepsilon_x = \frac{1}{E}[\sigma_x - v(\sigma_y + \sigma_z)]$$

$$\varepsilon_y = \frac{1}{E}[\sigma_y - v(\sigma_x + \sigma_z)]$$

$$\varepsilon_z = \frac{1}{E}[\sigma_z - v(\sigma_x + \sigma_y)] \qquad (3)$$

The method of superposition is applicable so long as the displacements are small and the corresponding strains do not substantially affect the action of the applied stresses. From Equations (3) it is seen that the relationships between compressive (or tensile) stresses and their corresponding strains are completely defined in terms of the two constants E and v.

RELATIONSHIP BETWEEN SHEAR STRESSES AND STRAINS

The ratio of shear stress to shear strain is known as the Rigidity Modulus, or the Modulus of Elasticity in shear, and is given the symbol G.

$$G = \frac{\tau}{\gamma}$$

Consider the element in Figure 6.2 subjected to the following stress system: σ_x, $\sigma_y = -\sigma_x$, $\sigma_z = 0$ (as shown in Figure 6.3A)

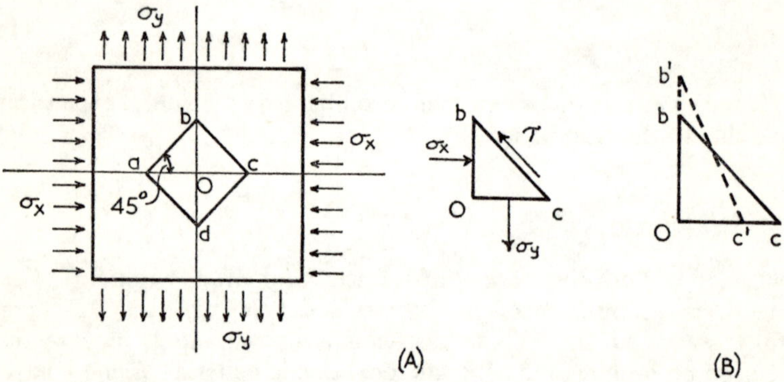

Fig. 6.3.

Consider an element abcd, with its plane perpendicular to the z axis and its sides at 45° to the x and y axes and examine the portion Obc of the element.

$$\text{Ob} = \text{Oc}, \quad \text{ab} = \frac{\text{Ob}}{\cos 45°}$$

If we assume that Ob = Oc = 1 then bc = 1/cos 45° and, assuming unit thickness parallel to the z axis:

$$\text{Stress on Ob} = \sigma_x \quad \therefore \text{ Force on Ob} = \sigma_x$$

$$\text{Stress on Oc} = \sigma_y \quad \therefore \text{ Force on Oc} = \sigma_y$$

Resolving forces perpendicular to bc:

$$\text{Component of } \sigma_x = \sigma_x \cos 45°; \quad \text{Component of } \sigma_y = \sigma_y \cos 45°$$

As these components act in opposite directions there can be no normal stress on bc. Resolving forces parallel to bc:

$$\text{Component of } \sigma_x = \sigma_x \cos 45°; \quad \text{Component of } \sigma_y = \sigma_y \cos 45°$$

These two tangential components act in the same direction. Hence the tangential force on bc is $2\sigma_x \cos 45°$.

Therefore the shear stress on bc, τ, is

$$\frac{2\sigma_x \cos 45°}{1/\cos 45°} = \sigma_x$$

It can be shown in a similar manner that the faces cd, ad and ab of the element are also only subjected to shear stresses. The element abcd is said to be in a state of pure shear. The elongation of bd is equal to the shortening of ac. The angle between ab and bc will change but, if we ignore the second order strains along ab and bc, the lengths of ab and bc will remain the same.

Considering our triangular element Obc further: Let b′ and c′ be the positions of b and c after the application of the stress system (Figure 6.3B):

The shear strain is the angular rotation of the element, i.e. the change in angle abc (= ψ say). If we consider angle Obc then the change in this angle will be $\psi/2$. The final value of angle Obc will be

$$\left(\frac{\pi}{4} - \frac{\psi}{2}\right) = \left(\frac{\pi}{4} + \frac{\gamma}{2}\right)$$

where γ is the shear strain and includes the negative sign. The final value of angle Obc is Ob′c′.

Hence

$$\frac{Oc'}{Ob'} = \tan\left(\frac{\pi}{4} + \frac{\gamma}{2}\right) = \frac{Oc - cc'}{Ob + bb'} = \frac{Oc + Oc\varepsilon_x}{Ob + Ob\varepsilon_y}$$

where the symbols ε_x and ε_y include the relevant signs.

Hence

$$\tan\left(\frac{\pi}{4} + \frac{\gamma}{2}\right) = \frac{1 + \varepsilon_x}{1 + \varepsilon_y} \quad \text{(As Ob = Oc)}$$

Now, from Equations (3) we can write down expressions for ε_x and ε_y:

$$\varepsilon_x = \frac{1}{E}(\sigma_x - v\sigma_y) = \frac{1}{E}(\sigma_x + v\sigma_x) = \frac{1 + v}{E}\sigma_x$$

$$\varepsilon_y = \frac{1}{E}(\sigma_y - v\sigma_x) = \frac{1}{E}(-\sigma_x - v\sigma_x) = -\frac{1 + v}{E}\sigma_x$$

$$\tan\left(\frac{\pi}{4} + \frac{\gamma}{2}\right) = \frac{\tan(\pi/4) + \tan(\gamma/2)}{1 - \tan(\pi/4)\tan(\gamma/2)}$$

$$= \frac{1 + \tan(\gamma/2)}{1 - \tan(\gamma/2)} = \frac{1 + (\gamma/2)}{1 - (\gamma/2)} \quad \text{(if } \gamma \text{ is small)}$$

Hence

$$\frac{1 + (\gamma/2)}{1 - (\gamma/2)} = \frac{1 + \varepsilon_x}{1 + \varepsilon_y}$$

or

$$1 + \varepsilon_y + \frac{\gamma}{2} = 1 + \varepsilon_x - \frac{\gamma}{2}$$

(ignoring second-order strains) or

$$\gamma = \varepsilon_x - \varepsilon_y$$

hence

$$\gamma = \frac{2(1 + v)}{E} \sigma_x = \frac{2(1 + v)}{E} \tau$$

hence

$$\frac{\tau}{\gamma} = \frac{E}{2(1 + v)} = G$$

The relationships between shearing stresses and their corresponding strains are completely defined in terms of the two constants E and v. The expression that has been evolved for γ gives the angular distortion between the (x, z) and (y, z) planes. If shearing stresses act on the sides of the element then the angular distortion between any two planes is determined from the shear stress components that act parallel to the two coordinate directions involved.

Hence

$$\gamma_{xy} = \frac{1}{G} \tau_{xy}; \quad \gamma_{yz} = \frac{1}{G} \tau_{yz}; \quad \gamma_{zx} = \frac{1}{G} \tau_{zx} \tag{4}$$

The compressions in Equations (3) and the angular distortions in Equations (4) are independent of one another and the general case of strains produced by three normal and three shear-stress components can be obtained by superposition.

VOLUMETRIC STRAIN

If we consider a unit cube, subjected to compressive principal stresses, σ_x, σ_y and σ_z then: new length in the x direction $= 1 - \varepsilon_x$, new length in the y direction $= 1 - \varepsilon_y$, and new length in the z direction $= 1 - \varepsilon_z$. Hence

$$\text{final volume} = (1 - \varepsilon_x)(1 - \varepsilon_y)(1 - \varepsilon_z)$$
$$= 1 - (\varepsilon_x + \varepsilon_y + \varepsilon_z)$$

ignoring second-order terms. Hence,

$$\text{change in volume} = 1 - \{1 - (\varepsilon_x + \varepsilon_y + \varepsilon_z)\}$$
$$= \varepsilon_x + \varepsilon_y + \varepsilon_z$$

Therefore the volumetric strain of the cube,

$$\frac{\text{change in volume}}{\text{original volume}} = \frac{\Delta V}{V} = \frac{\varepsilon_x + \varepsilon_y + \varepsilon_z}{1} = \varepsilon_x + \varepsilon_y + \varepsilon_z$$

And, expressing in terms of the applied stresses (from Equations (3)):

$$\frac{\Delta V}{V} = \frac{1 - 2v}{E}(\sigma_x + \sigma_y + \sigma_z)$$

The same volumetric strain will be experienced by the cube if it is subjected to a hydrostatic stress, σ_a, equal to the average of the applied stresses, i.e. if $\sigma_a = \frac{1}{3}(\sigma_x + \sigma_y + \sigma_z)$.

$$\frac{\Delta V}{V} = \frac{3(1 - 2v)}{E}\frac{(\sigma_x + \sigma_y + \sigma_z)}{3}$$
$$= \frac{3(1 - 2v)}{E}\sigma_a = \frac{\sigma_a}{K}$$

Where $K = E/3(1 - 2v)$ and is known as the Modulus of Volume Compressibility or the Bulk Modulus.

RELATIONSHIP BETWEEN VOLUMETRIC STRAIN AND STRESS

Equations (3) and (4) give the components of strain as functions of the components of stress. Sometimes it is useful to express the stress components as functions of strain.

Let $e = \varepsilon_x + \varepsilon_y + \varepsilon_z$ and $\theta = \sigma_x + \sigma_y + \sigma_z$. Then, adding Equations (3) together:

$$e = \frac{1 - 2v}{E}\theta \qquad (5)$$

In the case of a uniform hydrostatic pressure of value σ_a we have

$$\frac{\sigma_x + \sigma_y + \sigma_z}{3} = \sigma_a = \frac{\theta}{3}$$

and Equation (5) yields: $e = [3(1 - 2v)/E]\sigma_a$ giving the relationship between unit volume compression, e, and hydrostatic pressure, σ_a.

If we now solve Equations (3) for σ_x, σ_y and σ_z we obtain:

$$\varepsilon_x = \frac{1}{E}[\sigma_x - v(\sigma_y + \sigma_z)]$$

$$= \frac{1}{E}(\sigma_x + v\sigma_x - v\theta)$$

$$= \frac{1}{E}[(1 + v)\sigma_x - v\theta]$$

$$\therefore \quad \sigma_x = \frac{E\varepsilon_x}{1 + v} + \frac{v\theta}{1 + v}$$

$$= \frac{E}{1 + v}\varepsilon_x + \frac{vE}{(1 + v)(1 - 2v)} \cdot e$$

Using the notation

$$\lambda = \frac{vE}{(1 + v)(1 - 2v)} \quad \text{and} \quad G = \frac{E}{2(1 + v)}$$

the expressions for σ_x, σ_y and σ_z become:

$$\sigma_x = \lambda e + 2G\varepsilon_x$$

$$\sigma_y = \lambda e + 2G\varepsilon_y$$

$$\sigma_z = \lambda e + 2G\varepsilon_z \qquad (6)$$

PLANE STRESS

Although discussed in Chapter 5 plane stress, for the sake of completeness, is included in this section using a slightly different argument. If a thin plate is loaded along its boundary by forces applied parallel to the plane of the plate and distributed uniformly over its thickness the stress components σ_z, τ_{xz} and τ_{yz} are zero on both surfaces of the plate and we may assume that they are also zero within the plate (Figure 6.4). The state of stress is then specified by σ_x, σ_y and τ_{xy} and is known as the plane stress state. With plane stress the assumption is made that σ_x, σ_y and τ_{xy} do not vary throughout the thickness of the plate, i.e. they are assumed to be functions of x and y only.

It is also assumed that there will only be strains corresponding to the stress components, i.e. strains are ε_x, ε_y and γ_{xy}.

From Equation (3):

$$\varepsilon_x = \frac{1}{E}(\sigma_x - v\sigma_y)$$

$$\varepsilon_y = \frac{1}{E}(\sigma_y - v\sigma_x)$$

i.e.

$$\varepsilon_x = \frac{1}{E}[\sigma_x - v(E\varepsilon_y + v\sigma_x)]$$

$$\therefore \quad \sigma_x = \frac{E}{1 - v^2}(\varepsilon_x + v\varepsilon_y)$$

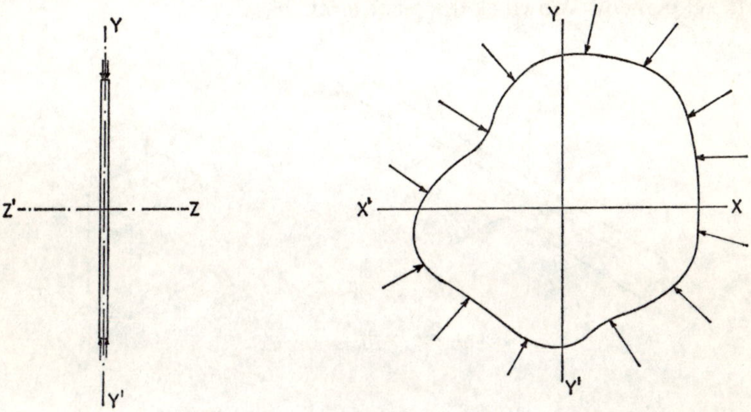

Fig. 6.4.

Similarly

$$\sigma_y = \frac{E}{1 - v^2}(\varepsilon_y + v\varepsilon_x)$$

From Equation (4):

$$\tau_{xy} = G\gamma_{xy} = \frac{E}{2(1 + v)}\gamma_{xy}$$

Hence

$$\tau_{xy} = \frac{E(1 - v)}{2(1 - v^2)}\gamma_{xy}$$

The stress–strain relationships for plane stress can therefore be expressed in matrix form:

$$\begin{bmatrix} \sigma_x \\ \sigma_y \\ \tau_{xy} \end{bmatrix} = \frac{E}{1 - v^2}\begin{bmatrix} 1 & v & 0 \\ v & 1 & 0 \\ 0 & 0 & \frac{1 - v}{2} \end{bmatrix}\begin{bmatrix} \varepsilon_x \\ \varepsilon_y \\ \gamma_{xy} \end{bmatrix}$$

PLANE STRAIN

When the dimension z of a body is very large we have the opposite extreme to plane stress. If z is large then we can assume that the end sections are confined between two smooth and rigid plates so that the displacement along the z direction is prevented. (See Figure 6.5.)

By symmetry there can be no displacement in the z direction at the centre and, as we have made sure that there are no displacements in the same direction at the ends it is reasonable to assume that, throughout the length of the body in the z direction, there are no displacements. This condition, where strains can only occur in the (x, y) plane, is known as the plane strain case.

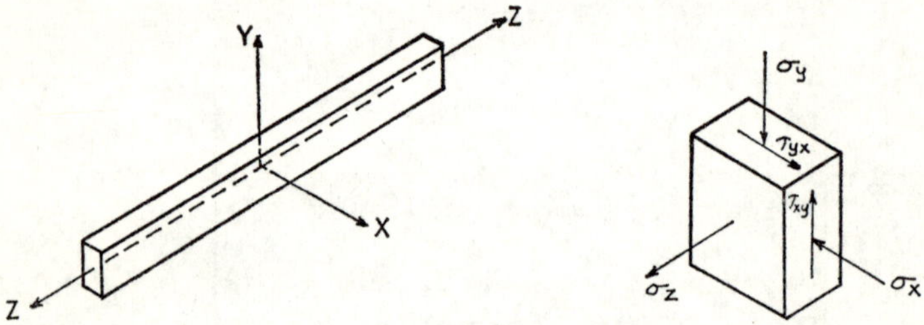

Fig. 6.5. Plane strain.

Consider two sections a unit distance apart: The displacement w in direction z is zero, i.e. $\varepsilon_z = 0$ and u and v are functions of x and y only. Hence $\gamma_{yz} = \gamma_{xz} = 0$. With this information the value of the longitudinal stress, σ_z, can be found from Equation (3)

$$\varepsilon_z = \frac{1}{E}[\sigma_z - \nu(\sigma_x + \sigma_y)] = 0$$

$$\therefore \quad \sigma_z = \nu(\sigma_x + \sigma_y)$$

Now, as the shear strains in the z direction are zero it is seen, from Equations (4) that the shear stresses τ_{zx} and τ_{yz} are also zero. The plane strain problem therefore reduces to the same form as the plane stress problem in that all that is required is to determine σ_x, σ_y and τ_{xy}.

From Equation (3):

$$\varepsilon_x = \frac{1}{E}[\sigma_x - \nu(\sigma_y + \sigma_z)]$$

Now

$$\sigma_z = v(\sigma_x + \sigma_y)$$

$$\therefore \quad \varepsilon_x = \frac{1}{E}[\sigma_x - v\sigma_y - v^2(\sigma_x + \sigma_y)]$$

$$= \frac{1}{E}[\sigma_x(1 - v^2) - v\sigma_y(1 + v)] \qquad \text{(A)}$$

Similarly

$$\varepsilon_y = \frac{1}{E}[\sigma_y(1 - v^2) - v\sigma_x(1 + v)]$$

i.e.

$$\sigma_y = \frac{E\varepsilon_y + v\sigma_x(1 + v)}{1 - v^2}$$

Substituting for σ_y in Equation (A) eventually leads to the expression:

$$\sigma_x = \frac{E(1 - v)}{(1 + v)(1 - 2v)}\left[\varepsilon_x + \frac{v}{1 - v}\varepsilon_y\right]$$

A similar expression can be evolved for σ_y.

τ_{xy}, as for plane stress, is given by the expression:

$$\tau_{xy} = \frac{E}{2(1 + v)}\gamma_{xy}$$

Hence the elasticity matrix for plane strain can be written as:

$$\begin{bmatrix} \sigma_x \\ \sigma_y \\ \tau_{xy} \end{bmatrix} = \frac{E(1 - v)}{(1 + v)(1 - 2v)} \begin{bmatrix} 1 & \frac{v}{1 - v} & 0 \\ \frac{v}{1 - v} & 1 & 0 \\ 0 & 0 & \frac{1 - 2v}{2(1 - v)} \end{bmatrix} \begin{bmatrix} \varepsilon_x \\ \varepsilon_y \\ \gamma_{xy} \end{bmatrix}$$

An important point about this expression is that it becomes inoperable for incompressible materials in which $v = 0{\cdot}5$. This is of little consequence with metals but in soils a saturated clay, in its initial loading stages, is often assumed to have $v = 0{\cdot}5$. Obviously the assumption of $v = 0{\cdot}5$ cannot be used in plane strain elasticity problems. The solution is to assume that $v = 0{\cdot}45$ or some other value near to $0{\cdot}5$, a realistic assumption when one considers the imponderables that apply in soils work.

STRAIN ENERGY

If we consider a unit cube, subjected to compressive principal stresses, σ_x, σ_y and σ_z acting on its faces then the work done on the cube by the resulting deformations will be:

$$U_t = \tfrac{1}{2}\sigma_x\varepsilon_x + \tfrac{1}{2}\sigma_y\varepsilon_y + \tfrac{1}{2}\sigma_z\varepsilon_z$$

Now

$$\varepsilon_x = \frac{1}{E}[\sigma_x - v(\sigma_y + \sigma_z)] \quad \text{etc.}$$

Hence

$$U_t = \frac{1}{2E}[\sigma_x^2 + \sigma_y^2 + \sigma_z^2 - 2v(\sigma_x\sigma_y + \sigma_y\sigma_z + \sigma_z\sigma_x)]$$

which is the expression for the total elastic energy stored in the cube. Generally the total strain energy will consist of two components, the strain energy due to the volumetric strain, U_v, and the strain energy due to the shear stress distortion, U_γ.
i.e.

$$U_t = U_v + U_\gamma$$

As shown earlier:

$$\frac{\Delta V}{V} = \frac{3(1-2v)}{E}\sigma_a$$

$$\therefore \quad U_v = \frac{3(1-2v)}{E}\sigma_a\frac{\sigma_a}{2}$$

$$= \frac{3(1-2v)}{2E}\sigma_a^2 = \frac{3(1-2v)}{2E}\frac{(\sigma_x + \sigma_y + \sigma_z)^2}{3}$$

$$= \frac{1-2v}{6E}(\sigma_x + \sigma_y + \sigma_z)^2$$

If we subtract the hydrostatic stress system (σ_a, σ_a, σ_a) from the applied general stress system (σ_x, σ_y, σ_z) the remaining part is the shear or deviatoric stress system, generally given the symbols (σ_x', σ_y', σ_z') or (s_x, s_y, s_z). In soil mechanics it is best to adopt the latter notation to avoid risk of confusion with effective stress symbols.

$$s_x = \sigma_x - \sigma_a = \frac{1}{3}(2\sigma_x - \sigma_y - \sigma_z) = \frac{2}{3}\left(\sigma_x - \frac{\sigma_y}{2} - \frac{\sigma_z}{2}\right)$$

$$s_y = \frac{2}{3}\left(\sigma_y - \frac{\sigma_z}{2} - \frac{\sigma_x}{2}\right); \qquad s_z = \frac{2}{3}\left(\sigma_z - \frac{\sigma_x}{2} - \frac{\sigma_y}{2}\right)$$

The work done by this deviatoric stress system is equal to the strain energy U_γ

$$U_\gamma = \frac{s_x \varepsilon_x'}{2} + \frac{s_y \varepsilon_y'}{2} + \frac{s_z \varepsilon_z'}{2}$$

where ε_x', ε_y' and ε_z' are the strains caused by the stress system.

Now

$$\begin{aligned}
\varepsilon_x' &= \frac{1}{E}[s_x - v(s_y + s_z)] \\
&= \frac{2}{3E}\left(\sigma_x - \frac{\sigma_y}{2} - \frac{\sigma_z}{2}\right) + \frac{2v}{3E}\left(\sigma_x - \frac{\sigma_y}{2} - \frac{\sigma_z}{2}\right) \\
&= \frac{2}{3E}(1 + v)\left(\sigma_x - \frac{\sigma_y}{2} - \frac{\sigma_z}{2}\right) \\
&= \frac{2}{3}\frac{(1 + v)}{E}\frac{3}{2}s_x = \frac{1 + v}{E}s_x = \frac{s_x}{2G}
\end{aligned}$$

similarly for ε_y' and ε_z'.

i.e.

$$\begin{aligned}
U_\gamma &= \frac{1}{4G}(s_x^{\,2} + s_y^{\,2} + s_z^{\,2}) \\
&= \frac{1}{6G}[(\sigma_x^{\,2} + \sigma_y^{\,2} + \sigma_z^{\,2}) - (\sigma_x\sigma_y + \sigma_y\sigma_z + \sigma_z\sigma_x)]
\end{aligned}$$

PLASTIC BEHAVIOUR

We shall commence our discussion of this section by considering the techniques employed for ductile materials such as metals.

PLASTIC YIELD THEORIES

A plastic yield theory must lead to a mathematical expression which can be used to determine what value of a complex stress system will cause a body to yield plastically, i.e. to reach its yield point. It has been shown that a complex stress system can be split into two components, a hydrostatic stress, σ_a, and a deviatoric stress system, s_x, s_y and s_z.

There are therefore two stress–strain aspects to consider:

(i) Volume changes due to σ_a

(ii) Distortional deformations due to the shear stresses produced by s_x, s_y and s_z.

From extensive observations it has been found that a hydrostatic stress causes volume changes but does not cause any significant plastic yield. The volume of a perfectly plastic material remains constant under plastic deformation.

The mathematical requirements of any yield theory can be quite simply stated:

(i) Hydrostatic stress has no influence on yielding.

(ii) Yield must be independent of the axes chosen to define the system. From (ii) it follows that yielding must be a function of the stress invariants. From (i) it is seen that yielding is unaffected by the hydrostatic stress σ_a, which is equal to $J_1/3$. Hence yielding can only depend upon the deviatoric or reduced stress invariants, in other words the criterion of yield must be some function of the deviatoric invariants J'_2 and J'_3.

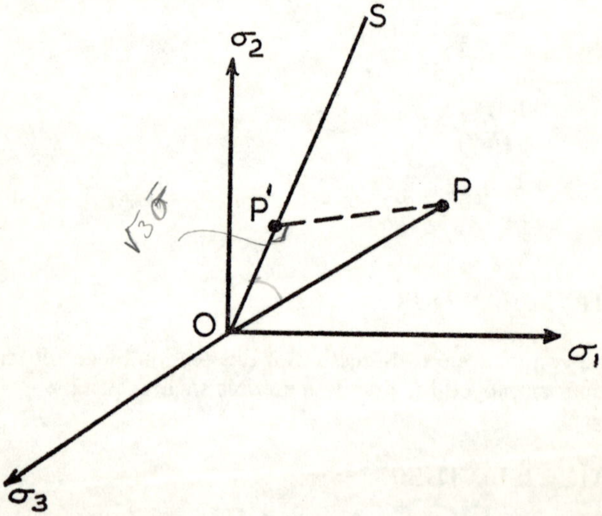

Fig. 6.6.

A state of general stress $(\sigma_1, \sigma_2, \sigma_3)$ is represented by point P in Figure 6.6. The direction cosines of OP are l, m and n. In the figure the direction OS represents the space diagonal with direction cosines $1/\sqrt{3}$.

The point P′ represents the hydrostatic stress system $(\sigma_a, \sigma_a, \sigma_a)$, where $\sigma_a = \frac{1}{3}(\sigma_1 + \sigma_2 + \sigma_3)$.

The length of $OP' = \sqrt{(\sigma_a^2 + \sigma_a^2 + \sigma_a^2)} = \sqrt{(3)}\sigma_a$. If an octahedral plane is now drawn through P′, normal to OS then the point P will lie on this plane. The proof of this statement can be illustrated as follows: If P does indeed lie on the octahedral plane then angle OP′P will be a right angle and

$$\cos \text{POP}' = \frac{\text{OP}'}{\text{OP}} = \frac{\sqrt{(3)}\sigma_a}{\text{OP}}$$

Now, from Chapter 2, $\cos \theta = l_1 l_2 + m_1 m_2 + n_1 n_2$.

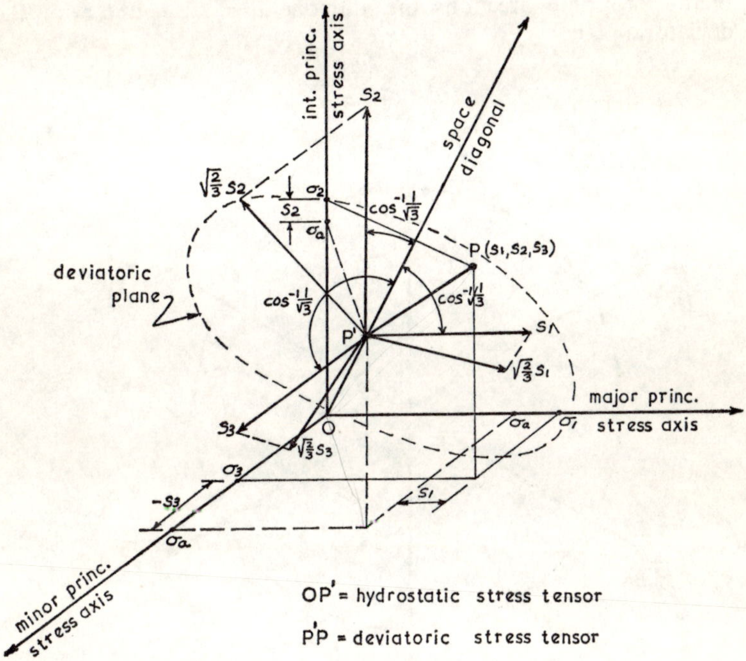

Fig. 6.7.

Hence

$$\cos \text{POP}' = \frac{1}{\sqrt{3}} l + \frac{1}{\sqrt{3}} m + \frac{1}{\sqrt{3}} n$$

$$= \frac{1 + m + n}{\sqrt{3}}$$

$$= \frac{\sigma_1/\text{OP} + \sigma_2/\text{OP} + \sigma_3/\text{OP}}{\sqrt{3}}$$

$$= \frac{3\sigma_a}{\text{OP}\sqrt{3}} = \frac{\sqrt{(3)}\sigma_a}{\text{OP}}$$

The two expressions agree, therefore P is in the octahedral plane passing through P′. The hydrostatic stress σ_a is taken as having no effect on plastic yielding which is therefore solely dependent upon the magnitude of the deviatoric stress tensor PP′. It is therefore convenient to consider the point P′ as a new origin and the plane containing P and P′, as it passes through the origin, is referred to as the deviatoric plane.

It then becomes possible to determine P by simply plotting, in their appropriate directions, the components of s_1, s_2 and s_3 on the deviatoric plane.

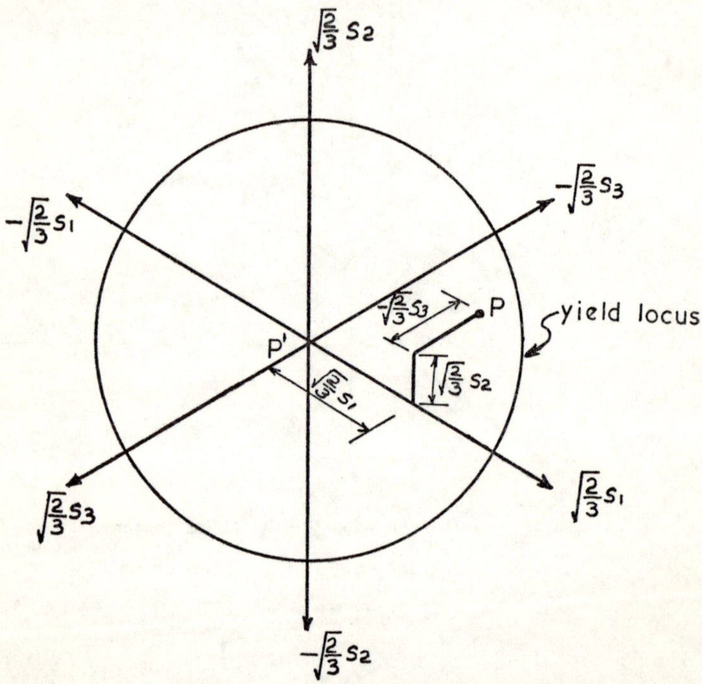

Fig. 6.8.

The directions of s_1, s_2 and s_3 are of course unchanged, they are all at an angle $\cos^{-1}\sqrt{(2/3)}$, to the deviatoric plane and they pass through P′.

The deviatoric plane is shown in Figure 6.7. Yielding will occur when the length PP′ reaches some limiting value. This limiting value is represented by the yield locus the boundaries of which are formed by the intersection of the yield surface and the particular deviatoric plane under consideration.

If we draw the deviatoric plane as if we were looking down the OP′ axis towards O then the deviatoric plane will lie in the plane of the paper and the principal deviatoric stress axes, $P's_1$, $P's_2$ and $P's_3$ will appear as if they are projected on to the deviatoric plane at 120° apart. The components of the deviatoric stresses are therefore plotted in the projected directions of $P's_1$, etc. to establish the point P (Figure 6.8). The projected lengths of s_1, s_2 and s_3 are $\sqrt{(2/3)}s_1$, $\sqrt{(2/3)}s_2$ and $\sqrt{(2/3)}s_3$ respectively.

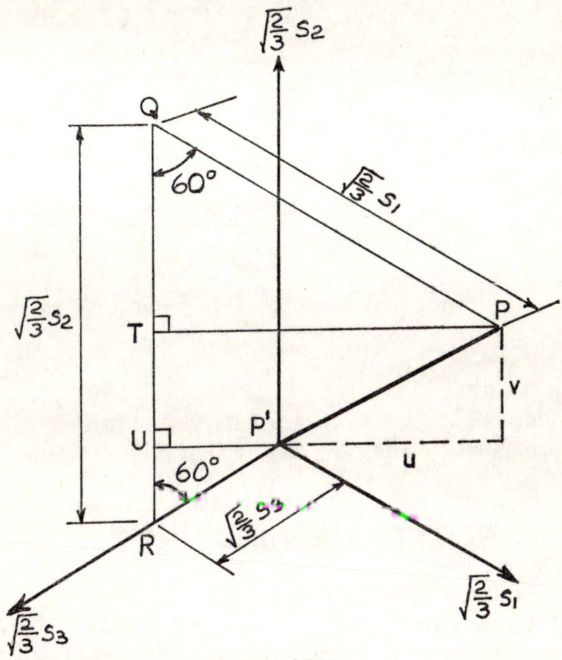

Fig. 6.9.

It is often convenient to draw the various stress components to represent the actual magnitude of their respective stresses. i.e. the component $\sqrt{(2/3)}s_1$ is plotted as s_1, etc., in effect then the various components are multiplied by $\sqrt{(2/3)}$.

Part of a deviatoric plane is shown in Figure 6.9.

In the figure it is seen that:

$$\sqrt{(2/3)}(s_1 - s_3) = PQ - P'R$$

$$= \frac{PT}{\sin 60°} - \frac{P'U}{\sin 60°} = \frac{2}{\sqrt{3}}u$$

Therefore

$$u = \frac{s_1 - s_3}{\sqrt{2}} = \frac{\sigma_1 - \sigma_3}{\sqrt{2}}$$

(as $s_1 = \sigma_1 - \sigma_a$ and $s_3 = \sigma_3 - \sigma_a$)

Also

$$\begin{aligned} \mathrm{v} = \mathrm{UT} &= \sqrt{(2/3)}s_2 - \mathrm{QT} - \mathrm{UR} \\ &= \sqrt{(2/3)}s_2 - \mathrm{QP}\cos 60° - \mathrm{P'R}\cos 60° \\ &= \frac{2s_2 - s_1 - s_3}{\sqrt{6}} = \frac{2\sigma_2 - \sigma_1 - \sigma_3}{\sqrt{6}} \end{aligned}$$

Hence

$$\begin{aligned} \mathrm{P'P} &= \sqrt{(u^2 + \mathrm{v}^2)} \\ &= \sqrt{\left\{\frac{(\sigma_1 - \sigma_3)^2}{2} - \frac{(2\sigma_2 - \sigma_1 - \sigma_3)^2}{6}\right\}} \\ &= \frac{1}{\sqrt{3}} \sqrt{\{(\sigma_1 - \sigma_2)^2 + (\sigma_2 - \sigma_3)^2 + (\sigma_3 - \sigma_1)^2\}} \\ &= \sqrt{(3)}\tau_{\mathrm{oct}} \end{aligned}$$

(The reader might like to check that, from Figure 6.6, the same result is obtained by using the relationship $\mathrm{P'P}^2 = \mathrm{OP}^2 - \mathrm{OP'}^2$.)

DETERMINATION OF THE YIELD LOCUS

Yield theories, for isotropic materials, are based on the assumption that failure (i.e. yield) occurs in a complex stress system when the maximum shear stress attained is related to the yield stress (or proof stress if no definite yield point) of the material when failed in pure tension or compression.

THE TRESCA OR MAXIMUM SHEAR STRESS THEORY

In this theory it is assumed that failure occurs when the maximum shear stress attained is equal to one half of the yield stress. A glance at the Mohr diagram of Figure 5.10 shows that the maximum shear stress in a complex system is $(\sigma_1 - \sigma_3)/2$. Hence the Tresca theory states that $(\sigma_1 - \sigma_3)/2 = Y/2$ or $(\sigma_1 - \sigma_3) = Y$, where $Y =$ yield stress.

THE VON MISES OR MAXIMUM SHEAR STRAIN ENERGY THEORY

Here it is assumed that failure will occur by plastic yield when the total shear strain energy per unit volume of the system is equal to the shear strain energy per unit volume at failure in pure tension. The shear strain energy is caused by the deviatoric stress system s_x, s_y and s_z. If the yield stress was Y and, assuming that it was applied in the x direction, then:

$$\sigma_x = Y; \quad \sigma_y = 0; \quad \sigma_z = 0$$

$$\sigma_a = \frac{\sigma_x + 0 + 0}{3} = \frac{\sigma_x}{3} = \frac{Y}{3}$$

Hence

$$s_x = Y - \frac{Y}{3} = \frac{2}{3}Y; \quad s_y = -\frac{Y}{3}; \quad s_z = -\frac{Y}{3}$$

Now

$$U_\gamma = \frac{1}{4G}(s_x^{\,2} + s_y^{\,2} + s_z^{\,2})$$

$$= \frac{1}{4G}\left(\frac{4Y^2}{9} + \frac{Y^2}{9} + \frac{Y^2}{9}\right)$$

$$= \frac{Y^2}{6G}$$

Writing σ_x, σ_y and σ_z as σ_1, σ_2 and σ_3 then the von Mises criterion is:

$$\frac{1}{6G}[\sigma_1^{\,2} + \sigma_2^{\,2} + \sigma_3^{\,2} - (\sigma_1\sigma_2 + \sigma_2\sigma_3 + \sigma_3\sigma_1)] = \frac{Y^2}{6G}$$

or

$$\sigma_1^{\,2} + \sigma_2^{\,2} + \sigma_3^{\,2} - (\sigma_1\sigma_2 + \sigma_2\sigma_3 + \sigma_3\sigma_1) = Y^2$$

The above equation can be expressed in terms of the octahedral shear stress:

$$\tau_{oct} = \frac{1}{3}\sqrt{\{(\sigma_1 - \sigma_2)^2 + (\sigma_2 - \sigma_3)^2 + (\sigma_3 - \sigma_1)^2\}}$$

$$= \frac{1}{3}\sqrt{\{2[\sigma_1^{\,2} + \sigma_2^{\,2} + \sigma_3^{\,2} - (\sigma_1\sigma_2 + \sigma_2\sigma_3 + \sigma_3\sigma_1)]\}}$$

$$\therefore \quad \tau_{oct}^{\,2} = \frac{2}{9}[\sigma_1^{\,2} + \sigma_2^{\,2} + \sigma_3^{\,2} - (\sigma_1\sigma_2 + \sigma_2\sigma_3 + \sigma_3\sigma_1)]$$

And the von Mises criterion becomes:

$$\frac{9}{2}\tau_{oct}^{\,2} = Y^2$$

or

$$\tau_{oct} = \frac{\sqrt{(2)}Y}{3}$$

THE MOHR–COULOMB THEORY

The Mohr strength theory is really an extension of the Tresca theory which, in turn, was probably based on Coulomb's work, hence the title. The theory assumes that the difference of the major and minor principal stresses is a function of their sum and can therefore be expressed as:

$$(\sigma_1 - \sigma_3) = f(\sigma_1 + \sigma_3)$$

As with the Tresca theory, any effect due to σ_2 is ignored.

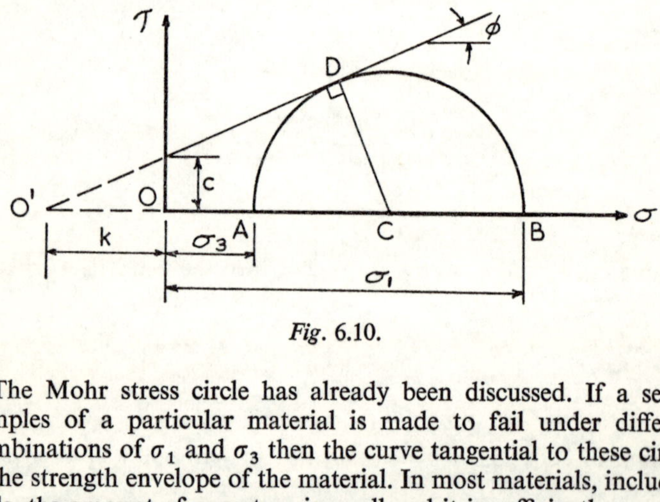

Fig. 6.10.

The Mohr stress circle has already been discussed. If a set of samples of a particular material is made to fail under different combinations of σ_1 and σ_3 then the curve tangential to these circles is the strength envelope of the material. In most materials, including soils, the amount of curvature is small and it is sufficiently accurate to assume that the strength envelope is a straight line. Figure 6.10 shows the positive half of one of these circles and the strength envelope.

The slope of the strength envelope from the normal stress axis is called the angle of internal friction of the material, φ. The intercept on the shear axis is the intrinsic pressure, i.e. the strength of the material under zero normal stress. In soil mechanics this intercept is called cohesion and given the symbol c. In Figure 6.10:

$$\sin\varphi = \frac{DC}{O'C} = \frac{\frac{1}{2}(\sigma_1 - \sigma_3)}{k + \frac{1}{2}(\sigma_1 + \sigma_3)} = \frac{\sigma_1 - \sigma_3}{2k + \sigma_1 + \sigma_3}$$

Hence

$$\sigma_1 - \sigma_3 = 2k \sin \varphi + (\sigma_1 + \sigma_3) \sin \varphi$$

Now

$$k = c \cot \varphi$$

$$\therefore \quad (\sigma_1 - \sigma_3) = 2c \cos \varphi + (\sigma_1 + \sigma_3) \sin \varphi$$

which is the general form of the Mohr–Coulomb yield theory.

PICTORIAL REPRESENTATION OF YIELD THEORIES

From Figure 6.8 it was seen that the solution to a yield problem can be obtained by plotting the point $P(\sigma_1, \sigma_2, \sigma_3)$ on the deviatoric plane and observing whether or not it lies within the yield locus. Obviously the yield locus will vary slightly depending upon which theory was used to formulate it.

THE VON MISES THEORY

The von Mises criterion is that, at failure, $\tau_{oct} = \surd(2/3)Y$.

Now it has been shown that

$$P'P = \surd(3)\tau_{oct}$$

Therefore on the deviatoric plane:

$$P'P = \surd(2/3)Y$$

Hence a circle of centre P′ and radius $\surd(2/3)Y$ on the diagram represents the yield locus of this theory (Figure 6.11). As has been mentioned the actual stress values are often plotted on the deviatoric plane. When this is done the radius of the circle becomes $\surd(2/3)\surd(3/2)Y = Y$.

THE TRESCA THEORY

The criterion of this theory is that $\sigma_1 - \sigma_3 = Y$.

Now, from Figure 6.9 it was shown that

$$u = \frac{\sigma_1 - \sigma_3}{\surd 2}$$

Therefore on the deviatoric plane:

$$u = \frac{Y}{\surd 2}$$

Hence, from Figure 6.9, it is seen that the yield locus consists of a line parallel to the $\sqrt{(2/3)}\sigma_2$ axis and is at $+Y/\sqrt{2}$ horizontally from P′.

Similarly if σ_1 and σ_3 are both negative stresses the locus will be a vertical line at a horizontal distance of $-Y/\sqrt{2}$ from P′. To draw the full yield locus one must assume that the three axes marked as σ_1, σ_2 and σ_3 on Figure 6.11 can become either the σ_1 or the σ_3 axis. The result is a regular hexagon, shown in Figure 6.11. Note that if the actual magnitude of the deviatoric stresses are plotted then the dimension $Y/\sqrt{2}$ becomes $\sqrt{(3)}(Y/2)$.

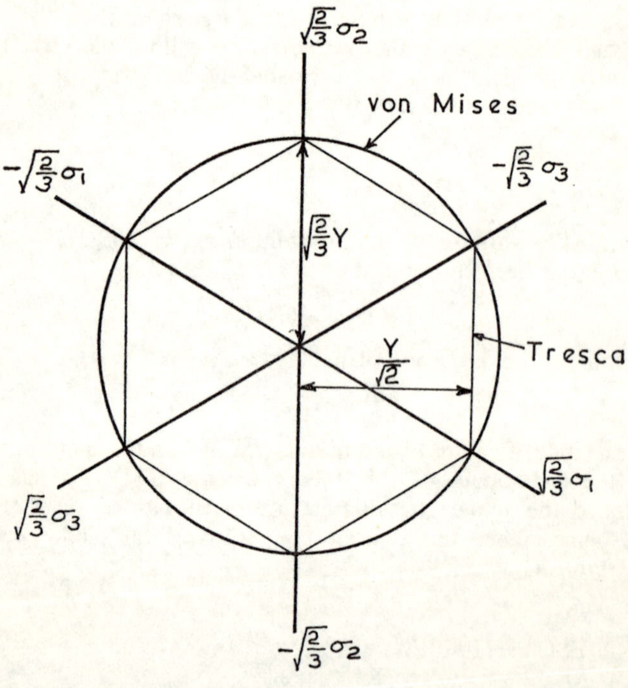

Fig. 6.11. *Yield loci for Tresca and von Mises theories (true dimensions).*

THE MOHR–COULOMB THEORY

Although relatively unimportant in metals this is the major theory used in soil mechanics. The yield locus of this theory, when drawn on to a deviatoric plane, is of the form shown in Figure 6.15B. Although the shape is constant the dimensions of the yield locus vary as, unlike the Tresca and von Mises theories, there is a dependence upon the hydrostatic stress. The yield surface is a flat sided conical figure varying in dimensions and reducing in size until it becomes a point when σ_1 and σ_3 are both tensile stresses of the same magnitude, i.e. at the point where $2c \cos \varphi - 2\sigma \sin \varphi = 0$, which is when $\sigma = c \cot \varphi$. In other words the apex of the yield surface is at a distance $\sqrt{(\sigma^2 + \sigma^2 + \sigma^2)} = \sqrt{(3)}c \cot \varphi$.

This is the characteristic difference between the Mohr–Coulomb theory and the Tresca and von Mises theories. The latter two have yield surfaces, both with the space diagonal as axis, but forming a hexagonal prism and a cylinder respectively.

YIELD IN SOILS

Any theory attempting to define the stress–strain relationships of a nonlinear material must be complex and particularly so when the material is soil.

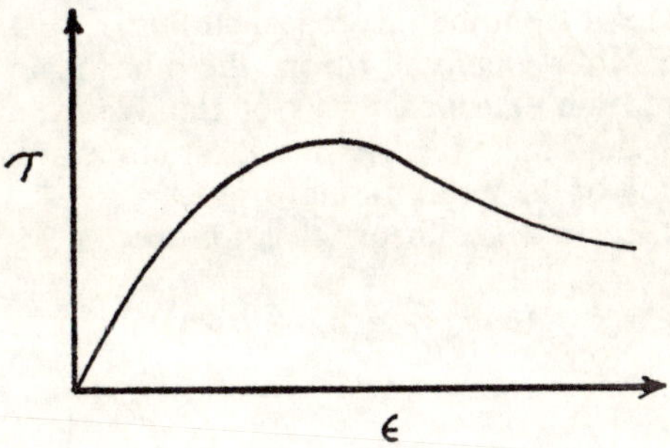

Fig. 6.12.

Unlike a perfect plastic the stress–strain relationships in a soil are not only time dependent but the angles of friction and cohesion do not remain constant and can experience considerable changes with strain. For example a dense sand dilates during shear and its angle of friction reduces with strain leading to a lower strength after the peak value has been reached (Figure 6.12).

Hence, in soils, principal stresses and axes can change both with strain and with time.

EFFECTIVE STRESSES

In most soil shear tests a measurement of the pore water pressure within the test sample is obtained. If principal stresses are expressed in terms of effective stress (total stress – pore water pressure) the effects of some variables are reduced. It is therefore often convenient to express stress analysis work in soils in terms of effective stresses. The strength envelopes of most soils when expressed in this form, have a small, if not negligible, cohesive intercept and the soil can be considered as if it were cohesionless. For simplicity only this case will be considered from now on.

The Mohr–Coulomb criterion therefore becomes:

$$(\sigma_1 - \sigma_3) = (\sigma_1 + \sigma_3) \sin \varphi$$

YIELD THEORIES IN SOIL MECHANICS

In soils work it is, and has been, the general practice to use the Mohr–Coulomb yield theory in order to interpret shear test results. However, due to the many variables involved, the actual yield surface can depart from the theoretical surface.

Besides the Mohr–Coulomb theory the other main theories that have been used are extended forms of the Tresca and von Mises theories. With both these theories the shear stress at failure is taken to be a function of the mean normal stress.

The extended von Mises theory therefore becomes:

$$\sigma_1^2 + \sigma_2^2 + \sigma_3^2 - (\sigma_1\sigma_2 + \sigma_2\sigma_3 + \sigma_3\sigma_1) = k_1^2 \left(\frac{\sigma_1 + \sigma_2 + \sigma_3}{3}\right)^2$$

or

$$\tau_{oct} = \frac{\sqrt{(2)}}{3} k_1 \left(\frac{\sigma_1 + \sigma_3 + \sigma_2}{3}\right)$$

and the extended Tresca theory becomes:

$$\sigma_1 - \sigma_3 = k_2 \left(\frac{\sigma_1 + \sigma_2 + \sigma_3}{3}\right)$$

Bishop (1966) has shown that, at the present stage in development, both the extended Tresca and von Mises theories can lead to an erroneous overestimation of strength in many cases leading to the conclusion that the Mohr–Coulomb criterion is the only simple theory that can consistently give safe results. The reader is referred to Bishop's paper for a fuller discussion.

THE TRIAXIAL TEST

In the normal triaxial test the test specimen is enclosed in a rubber membrane, inserted in a perspex cell, subjected to an all round water pressure and failed under an increasing axial stress. This is known as a compression test (in which $\sigma_1 > \sigma_3 = \sigma_2$). The Mohr–Coulomb theory can therefore be expressed as:

$$\sigma_1 - \sigma_3 = \sigma_1 \sin \varphi + \sigma_3 \sin \varphi$$

i.e.

$$\sigma_{1f}(\max) = \sigma_{3f} \frac{1 + \sin \varphi}{1 - \sin \varphi}$$

where σ_{1f} and σ_{3f} are the respective stresses at failure.

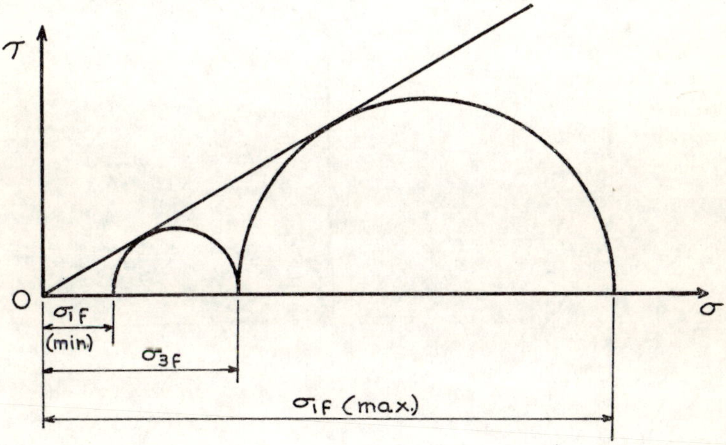

Fig. 6.13.

It is possible to fail the sample in axial tension by first subjecting it to equal pressures σ_1 and σ_3 and then gradually reducing σ_1 below the value of σ_3 until failure occurs. This test is known as the axial extension test.

In this case the Mohr–Coulomb criterion becomes:

$$\sigma_{1f}(\min) = \sigma_{3f} \frac{1 - \sin \varphi}{1 + \sin \varphi} \qquad (\text{where } \sigma_1 < \sigma_2 = \sigma_3)$$

The Mohr circle diagram illustrating the maximum and minimum values of σ_1 for a fixed value of σ_3 is shown in Figure 6.13. Strictly speaking, as there is a reversal of stresses in the axial extension test there should also be an exchange of suffixes.

A stress state in which $\sigma_1 \gtrless \sigma_3 = \sigma_2$ can be shown in the plane σ_1OA of Figure 6.14A where OA is a line drawn in the σ_2, σ_3 plane such that all points along it satisfy the relationship $\sigma_2 = \sigma_3$. The line OA will have direction cosines $1/\sqrt{2}$ with both the σ_2 and the σ_3 axis.

In Figure 6.14B is given a normal view of plane σ_1OA with OA drawn horizontal and Oσ_1 vertical. On this plane, at any point, $\sigma_2 = \sigma_3$ and hence this plane must contain the space diagonal which can be plotted on it. On this diagram the stress values of σ_1 will appear to a normal scale but as the diagram only shows the projected values of σ_2 and σ_3 these two values will scale $\sqrt{(2)}\sigma_2$ and $\sqrt{(2)}\sigma_3$, equal to each other.

If we assume that φ remains constant then we can draw the two boundaries for the Mohr–Coulomb theory on the diagram.

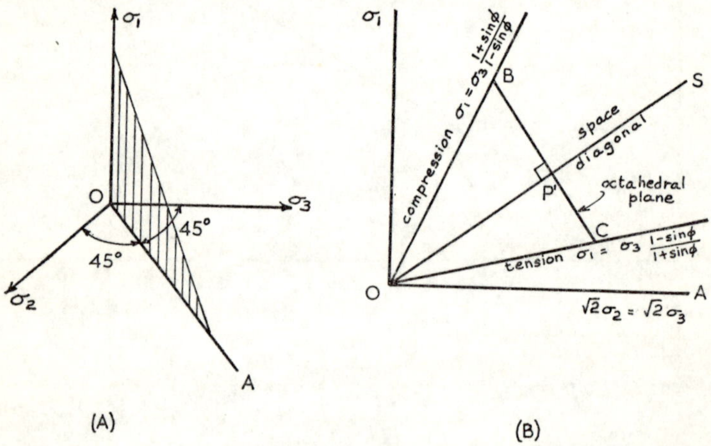

Fig. 6.14.

For axial compression the boundary has the equation:

$$\sigma_1 = \sigma_3 \frac{1 + \sin\varphi}{1 - \sin\varphi}$$

and for axial extension:

$$\sigma_1 = \sigma_3 \frac{1 - \sin\varphi}{1 + \sin\varphi}$$

If we select a point B on either of the boundaries then the line from B, at right angles to the space diagonal, will be the octahedral plane corresponding to the stress system at B and will locate the point representing the hydrostatic stress system, P′.

The portion of the octahedral plane that is enclosed within the boundaries of the Mohr–Coulomb theory is shown in Figure 6.15. We have already established the points B and C on the σ_1 and $\sigma_2 = \sigma_3$ axes. By working around the diagram, first substituting σ_1 for σ_2 and $(\sigma_2 = \sigma_3)$ for $(\sigma_1 = \sigma_3)$, etc. we can establish the three B and the three C positions. The form of the Mohr–Coulomb yield surface is therefore seen to be an irregular hexagon (Figure 6.15B). The yield surface is a flat sided conical figure, reducing to a point at the origin of the stress system, unless cohesion has been included, in which case the apex is at some distance below the origin on the space diagonal.

Example

A cohesionless soil sample failed in axial compression at a major principal stress of 10. Given that $\varphi = 30°$ draw the yield surface of the Mohr–Coulomb theory as viewed on the corresponding octahedral plane. The procedure is to prepare the diagram of the σ_1 OA plane and to mark on it the position of the two Mohr–Coulomb boundaries.

$$\varphi = 30° \qquad \therefore \quad \frac{1 + \sin\varphi}{1 - \sin\varphi} = 3; \qquad \frac{1 - \sin\varphi}{1 + \sin\varphi} = \frac{1}{3}$$

Select some arbitrary value for σ_3 and determine the corresponding values for σ_1 on the two boundaries. Plot the two points $(\sigma_{1f}(\text{max}), \sigma_{3f})$ and $(\sigma_{1f}(\text{min}), \sigma_{3f})$. As the soil is cohesionless the boundaries will be the lines drawn from O which pass through these points.

Say $\sigma_3 = 6$. Then $\sigma_{1f}(\text{max}) = 3 \times 6 = 18$; $\sigma_{1f}(\text{min}) = \frac{1}{3} \times 6 = 2$ and $\sqrt{(2)}\sigma_3 = 1{\cdot}414 \times 6 = 8{\cdot}48$.

Hence the point F, on the compressive boundary, is $(\sigma_1 = 18, \sigma_3 = 8{\cdot}48)$ and the point D, on the extension boundary, is $(\sigma_1 = 2, \sigma_3 = 8{\cdot}48)$. The boundaries are shown drawn in Figure 6.15A. Having drawn the boundaries it is now necessary to establish the position of the space diagonal. Again, if we choose an arbitrary value, say the point $(\sigma_1 = 6, \sigma_2 = 6, \sigma_3 = 6)$ then we can establish the point E which will have coordinates $(\sigma_a, \sigma_a, \sigma_a)$, where $\sigma_a = 6$. The point E is represented on the digram by $(\sigma_1 = 6, \sqrt{(2)}\sigma_3 = 8{\cdot}48)$ The line OE gives the direction and position of the space diagonal OS. The line from B, at right angles to OS will cut OS at P′ and the extension failure boundary at C. The line BC therefore gives the extent of the octahedral plane between the two boundaries. If we plot the octahedral plane, as viewed along the space diagonal we can plot the distances P′B and P′C to give two points on the yield surface (Figure 6.15B). Moving round the diagram and considering

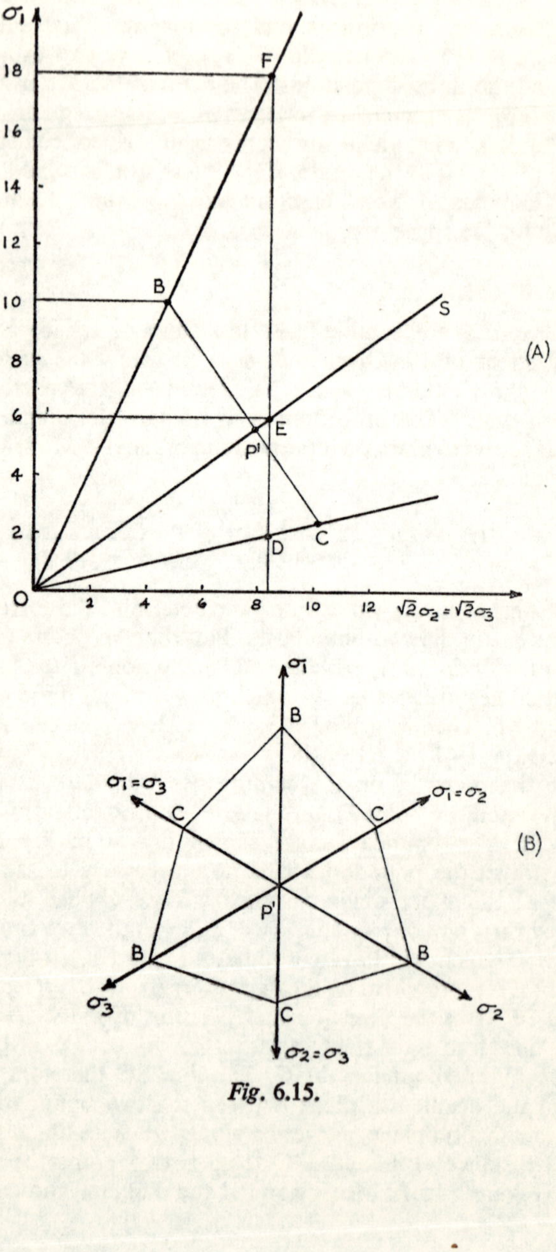
σ1
18
16
14
12
10
8
6
4
2
O
2
4
6
8
10
12
√2σ2 = √2σ3
F
B
S
(A)
E
P'
D
C
σ1
B
σ1 = σ3
σ1 = σ2
C
C
(B)
P'
B
B
σ3
C
σ2
σ2 = σ3

Fig. 6.15.

σ_2 to be σ_1 and ($\sigma_1 = \sigma_3$) to be ($\sigma_2 = \sigma_3$) we can plot P′B and P′C in these two directions and, finally, considering σ_3 to be σ_1 and ($\sigma_1 = \sigma_2$) to be ($\sigma_2 = \sigma_3$) the other two positions of B and C can be plotted. The lines joining these points gives the outline of the failure surface. As given in the question the point B is on the axial compression boundary surface where $\sigma_1 = 10$ (or where $\sigma_3 = 3{\cdot}33$).

For interest we can check on the coordinates of the point P′. The coordinates must be ($\sigma_a, \sigma_a, \sigma_a$) where

$$a = \left(\frac{10 + 3{\cdot}33 + 3{\cdot}33}{3}\right) = 5{\cdot}59$$

The diagram should therefore have point P′ at coordinates

$$(\sigma_1 = 5{\cdot}59, \quad \sqrt{(2)}\sigma_3 = \sqrt{(2)} \times 5{\cdot}59)$$

i.e. (5·59, 7·85) which is seen to be so.

7

The Finite Element Method of Stress Analysis

INTRODUCTION

IF a load W is applied at a point on the surface of a body then the displacement at another point on the surface is related to it, i.e.

$$\Delta = f(W)$$

If we have a system of loads, $W_1, W_2, \ldots, W_n$ applied to a body at points 1, 2, . . ., and n and if we assume that the system is linear and that the resulting deformations and strains are small we can apply the principle of superposition to determine the deformation at any one of the loaded points.

$$\Delta_i = f_{i1}W_1 + f_{i2}W_2 + \ldots + f_{in}W_n$$

where Δ_i is the deflection at the point i measured in the direction of W_i. Similar equations can be obtained for all the points and they can be expressed in matrix form:

$$\mathbf{\Delta} = \mathbf{FW}$$

where

$$\mathbf{\Delta} = \begin{bmatrix} \Delta_1 \\ \Delta_2 \\ \cdot \\ \cdot \\ \cdot \\ \Delta_n \end{bmatrix} \quad \mathbf{F} = \begin{bmatrix} f_{11} & f_{12} & f_{13} & \cdots & f_{1n} \\ f_{21} & f_{22} & f_{23} & \cdots & f_{2n} \\ \cdot & \cdot & \cdot & \cdots & \cdot \\ \cdot & \cdot & \cdot & \cdots & \cdot \\ \cdot & \cdot & \cdot & \cdots & \cdot \\ f_{n1} & f_{n2} & f_{n3} & \cdots & f_{nn} \end{bmatrix} \quad \mathbf{W} = \begin{bmatrix} W_1 \\ W_2 \\ \cdot \\ \cdot \\ \cdot \\ W_n \end{bmatrix}$$

where **F** is known as the flexibility matrix of the body. Maxwell's theorem of reciprocal deflections states that if a load W at a point 1 causes a deflection Δ at a point 2 then if the load W is applied at point 2 the deflection at point 1 will be Δ. Hence, in the above matrix $a_{ij} = a_{ji}$. The flexibility matrix is symmetric of dimensions $n \times n$. The elements of the matrix are often referred to as the flexibility coefficients. Obviously an inverse relationship must exist:

$$\mathbf{W} = \mathbf{F}^{-1}\mathbf{\Delta}$$
$$= \mathbf{K}\mathbf{\Delta}$$

K, the inverse of **F**, is known as the stiffness matrix and is also symmetrical of dimensions $n \times n$. Hence there are two ways of expressing the relationships between the displacements at a series of points and the loads applied at those points leading to two possible methods for solving stress problems:

$$\mathbf{\Delta} = \mathbf{F}\mathbf{W}$$

(the displacement or flexibility method) or

$$\mathbf{W} = \mathbf{K}\mathbf{\Delta}$$

(the force or stiffness method). With present day computer facilities it is the stiffness method that is generally used.

THE FINITE ELEMENT METHOD

The principle of the method can best be illustrated by considering a typical problem in plane stress. Figure 7.1A shows a rectangular

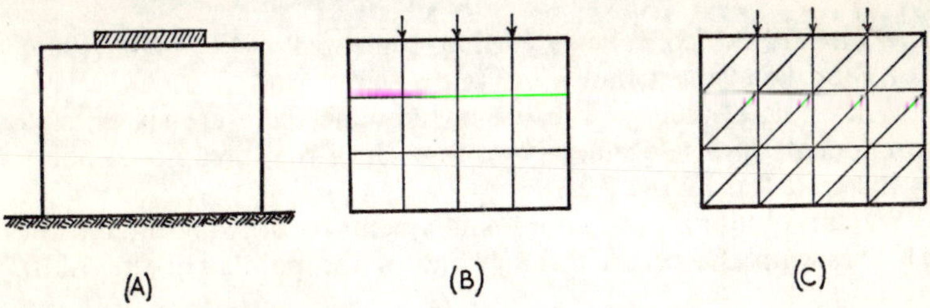

Fig. 7.1.

laminar subjected to a normal stress along part of its longer edge and it is required to determine the stress distribution within it.

The procedure is to split the rectangle into a series of elements, preferably all of the same type, which are physically separated from each other except at the nodal points. The applied stress can be considered to act as a set of concentrated loads applied at the relevant nodal points. In plane stress problems rectangular or triangular elements are generally used (Figures 7.1B and C). The continuum has now been given the form of a series of connected elements and can be regarded as a structure to which the normal rules of structural analysis can be applied.

The main feature of the evolved system is that, except at the nodal points where the elements are assumed joined, there is now no continuity and the new system appears to be much more flexible than the original. It may be stated however that if it is assumed that the stresses across each element vary linearly then the resulting strains are considerably limited and the edges of the elements tend to keep together and most of the continuity is retained.

It might seem that considerable practice is required to decide on a suitable arrangement of elements. This is not generally so as satisfactory results can be obtained with large elements although improvements are generally noted when a more refined grid, leading to smaller elements, is used.

Briefly then the new system consists of a set of elements, only connected at a small number of nodal points. As the material characteristics of each element are related to the forces and resulting deformations at the nodes it is possible to evaluate both the stiffness and the flexibility matrices either one of which can be used to solve the problem. We will consider the problem of evaluating the stiffness matrix first.

STIFFNESS MATRIX OF A TRIANGULAR ELEMENT

The method which follows is that suggested by Clough (1965). Consider the element shown in Figure 7.2A.

When this element is stressed in its plane there are six possible components of displacement, two at each of the three nodal points as illustrated in Figure 7.2A.

We can obtain a local coordinate system by drawing the $0X$ and $0Y$ axes with the origin through one of the points (Figure 7.2B).

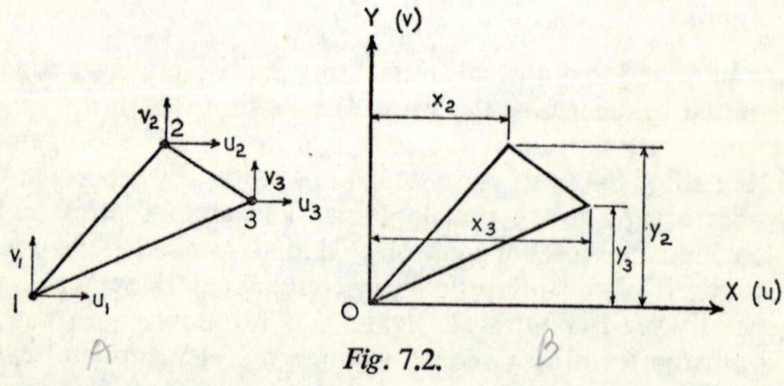

Fig. 7.2.

We must now turn our attention to the internal deformation that can take place within the element. There will be two components:

Let $u =$ the internal displacement component in the x direction.

Let $\mathrm{v} =$ the internal displacement component in the y direction.

Both u and v are assumed to vary linearly in the x and y directions respectively. i.e. u and v are functions of x and y and can therefore be expressed in terms of displacement functions of the form:

$$u = ax + by + c$$

$$\mathrm{v} = dx + ey + f$$

The two displacement functions can be written as

$$u = \alpha_1 + x\alpha_2 + y\alpha_3$$

$$\mathrm{v} = \alpha_4 + x\alpha_5 + y\alpha_6$$

or, in matrix form:

$$\begin{bmatrix} u \\ \cdot \\ \cdot \\ \cdot \\ \mathrm{v} \end{bmatrix} = \begin{bmatrix} 1 & x & y & 0 & 0 & 0 \\ \cdot & & & & & \cdot \\ \cdot & & & & & \cdot \\ \cdot & & & & & \cdot \\ 0 & 0 & 0 & 1 & x & y \end{bmatrix} \begin{bmatrix} \alpha_1 \\ \cdot \\ \cdot \\ \cdot \\ \alpha_6 \end{bmatrix}$$

As stated, there is a component of displacement in both the x and y directions at each nodal point. These six components can be evaluated from the general displacement functions just derived by substituting the actual coordinates of the nodes into the matrix:

$$\begin{bmatrix} u_1 \\ u_2 \\ u_3 \\ \mathrm{v}_1 \\ \mathrm{v}_2 \\ \mathrm{v}_3 \end{bmatrix} = \begin{bmatrix} 1 & 0 & 0 & 0 & 0 & 0 \\ 1 & x_2 & y_2 & 0 & 0 & 0 \\ 1 & x_3 & y_3 & 0 & 0 & 0 \\ 0 & 0 & 0 & 1 & 0 & 0 \\ 0 & 0 & 0 & 1 & x_2 & y_2 \\ 0 & 0 & 0 & 1 & x_3 & y_3 \end{bmatrix} \begin{bmatrix} \alpha_1 \\ \alpha_2 \\ \alpha_3 \\ \alpha_4 \\ \alpha_5 \\ \alpha_6 \end{bmatrix}$$

It is seen that as $x_1 = y_1 = 0$ $u_1 = \alpha_1$ and $\mathrm{v}_1 = \alpha_4$.

The above relationships are of the form:

$$\begin{bmatrix} u \\ \mathrm{v} \end{bmatrix} = \mathbf{A}\boldsymbol{\alpha}$$

which we will express as

$$\mathbf{u} = \mathbf{A}\boldsymbol{\alpha}$$

$\alpha_1, \alpha_2, \ldots, \alpha_6$ are of course the components of the displacement vector **u** related to some general system of coordinates. We will call α_1, α_2 etc. the generalised coordinate displacement components.

EVALUATION OF INTERNAL STRAINS

As was shown in Chapter 6:

$$\frac{\partial u}{\partial x} = \varepsilon_x; \quad \frac{\partial \mathrm{v}}{\partial y} = \varepsilon_y; \quad \gamma_{xy} = \frac{\partial u}{\partial y} + \frac{\partial \mathrm{v}}{\partial x}$$

If we return to our general displacement functions:

$$\begin{bmatrix} u \\ \mathrm{v} \end{bmatrix} = \begin{bmatrix} 1 & x & y & 0 & 0 & 0 \\ 0 & 0 & 0 & 1 & x & y \end{bmatrix} \begin{bmatrix} \alpha_1 \\ \cdot \\ \cdot \\ \cdot \\ \alpha_6 \end{bmatrix}$$

and differentiate accordingly we obtain the various strain components within the element. Hence:

$$\begin{bmatrix} \varepsilon_x \\ \varepsilon_y \\ \gamma_{xy} \end{bmatrix} = \begin{bmatrix} 0 & 1 & 0 & 0 & 0 & 0 \\ 0 & 0 & 0 & 0 & 0 & 1 \\ 0 & 0 & 1 & 0 & 1 & 0 \end{bmatrix} \begin{bmatrix} \alpha_1 \\ \alpha_2 \\ \alpha_3 \\ \alpha_4 \\ \alpha_5 \\ \alpha_6 \end{bmatrix}$$

which is of the form $\boldsymbol{\varepsilon} = \mathbf{B}\boldsymbol{\alpha}$.

EVALUATION OF INTERNAL STRESSES

The elastic relationship between stress and strain components for plane stress has been given in Chapter 6. In matrix form, the relationship is:

$$\begin{bmatrix} \sigma_x \\ \sigma_y \\ \tau_{xy} \end{bmatrix} = \frac{E}{1 - v^2} \begin{bmatrix} 1 & v & 0 \\ v & 1 & 0 \\ 0 & 0 & \frac{1 - v}{2} \end{bmatrix} \begin{bmatrix} \varepsilon_x \\ \varepsilon_y \\ \gamma_{xy} \end{bmatrix}$$

which is of the form:

$$\boldsymbol{\sigma} = \mathbf{D}\boldsymbol{\varepsilon}$$

It is at this point that one of the advantages of the finite element method is apparent. As will be shown later it is possible to insert a different form for matrix **D** when the stress–strain characteristics of the element are not linear. We have now obtained two expressions: Internal strains,

$$\boldsymbol{\varepsilon} = \mathbf{B}\boldsymbol{\alpha}$$

Internal stresses,

$$\boldsymbol{\sigma} = \mathbf{D}\boldsymbol{\varepsilon} = \mathbf{D}\mathbf{B}\boldsymbol{\alpha}$$

where

$$\mathbf{B} = \begin{bmatrix} 0 & 1 & 0 & 0 & 0 & 0 \\ 0 & 0 & 0 & 0 & 0 & 1 \\ 0 & 0 & 1 & 0 & 1 & 0 \end{bmatrix} \quad \mathbf{D} = \frac{E}{1 - v^2}\begin{bmatrix} 1 & v & 0 \\ v & 1 & 0 \\ 0 & 0 & \dfrac{1 - v}{2} \end{bmatrix} \quad \boldsymbol{\alpha} = \begin{bmatrix} \alpha_1 \\ \alpha_2 \\ \alpha_3 \\ \alpha_4 \\ \alpha_5 \\ \alpha_6 \end{bmatrix}$$

The stiffness matrix can now be derived in terms of the generalised coordinates by the principle of virtual work.

We can compound the internal stresses into a series of stress resultants, or forces, acting at the nodal points of the element in the directions of the generalised coordinate displacement components $\alpha_1 \ldots \alpha_6$. These stress resultants can be represented as

$$\mathbf{S} = \begin{bmatrix} S_1 \\ \cdot \\ \cdot \\ \cdot \\ S_6 \end{bmatrix}$$

and, by definition, are in equilibrium with the elemental stresses. If we now consider that this force system is externally applied and that each nodal point is successively subjected to a unit virtual displacement in the direction of a particular generalised coordinate displacement, i.e. $\alpha_1 = 1$, $\alpha_2 = 1$ etc., then we can obtain an expression for the external work, W_e, done on the element. (External work done) = (Displacements) × (Applied forces).

$$\mathbf{W}e = \bar{\boldsymbol{\alpha}}^t\mathbf{S}$$

Note that the horizontal bar above indicates that the displacement is virtual and also that, for multiplication the transpose of $\boldsymbol{\alpha}$ must be used. As the unit displacements are successively applied the matrix expressing nodal displacements is a unit matrix.
Hence the apparent column vector $\overline{\boldsymbol{\alpha}^t}$ is a unit matrix **I**.

$$W_e = \mathbf{IS} = \mathbf{S}$$

From Castigliano's theorem the external work done on the system must equal the internal work.

Internal work = (internal stress) × (internal strain) × (volume). To obtain an expression for the internal work we must consider an elemental volume, dV, of the total element and then integrate the resulting expression.

The internal stresses are σ as S was defined as the force system that is in equilibrium with the elemental stresses. Let the virtual strains caused by the virtual displacements be $\bar{\varepsilon}$. Internal work = dW_i

$$\begin{aligned} \mathrm{d}\mathbf{W}_i &= \overline{\boldsymbol{\varepsilon}^t}\boldsymbol{\sigma}\,\mathrm{d}V \\ &= \overline{\boldsymbol{\alpha}^t}\mathbf{B}^t\boldsymbol{\sigma}\,\mathrm{d}V \\ &= \overline{\boldsymbol{\alpha}^t}\mathbf{B}^t\mathbf{D}\mathbf{B}\boldsymbol{\alpha}\,\mathrm{d}V \end{aligned}$$

($\bar{\boldsymbol{\varepsilon}} = \mathbf{B}\bar{\boldsymbol{\alpha}}$. Hence $\overline{\boldsymbol{\varepsilon}^t} = \overline{\boldsymbol{\alpha}^t}\mathbf{B}^t$ see Chapter 3).
The total internal work done is the integration of the above expression over the whole volume. The expression can be written as:

$$\mathrm{d}\mathbf{W}_i = T\overline{\boldsymbol{\alpha}^t}\mathbf{B}^t\mathbf{D}\mathbf{B}\boldsymbol{\alpha}\,\mathrm{d}x\,\mathrm{d}y$$

where T = thickness of the element. Hence

$$\mathbf{W}_i = T[\textstyle\iint \mathbf{B}^t\mathbf{D}\mathbf{B}\,\mathrm{d}x\,\mathrm{d}y]\boldsymbol{\alpha}$$

(As $\overline{\boldsymbol{\alpha}^t} = \mathbf{I}$). T and $\boldsymbol{\alpha}$ can be taken outside the brackets as they are independent of x and y.

To carry out the integration the various matrices are first multiplied together to yield a single 6 × 6 matrix. Each term is then integrated within the limits of the problem.

Hence the final expression, in matrix form will be:

$$\mathbf{W}_i = \bar{\mathbf{k}}\boldsymbol{\alpha}$$

where

$$\bar{\mathbf{k}} = T[\textstyle\iint \mathbf{B}^t\mathbf{D}\mathbf{B}\,\mathrm{d}x\,\mathrm{d}y]$$

and, as

$$\mathbf{W}_e = \mathbf{W}_i$$

$$\mathbf{S} = \bar{\mathbf{k}}\boldsymbol{\alpha}$$

which is an expression of the form (Force) = $\bar{\mathbf{k}}$ × (displacement). Hence $\bar{\mathbf{k}}$ is the stiffness of the element expressed in terms of the general coordinate displacement vector $\boldsymbol{\alpha}$. $\bar{\mathbf{k}}$ must now be transformed to **k**, the stiffness of the element in terms of its nodal point coordinate displacements, u and v.

The relationship between the generalised and the local coordinates can be derived if we use the fact that the total strain energy stored in the system is the same no matter what coordinate axes are used i.e.

$$\tfrac{1}{2}(u_1 W_1 + \mathrm{v}_2 W_2 + u_3 W_3 + \ldots) = \tfrac{1}{2}(\alpha_1 S_1 + \alpha_2 S_2 + \alpha_3 S_3 + \ldots)$$

$$\therefore \quad \mathbf{u}^t\mathbf{W} = \boldsymbol{\alpha}^t\mathbf{S} = [\mathbf{A}^{-1}\mathbf{u}]^t\mathbf{S}$$

$$= \mathbf{u}^t[\mathbf{A}^{-1}]^t\mathbf{S}$$

$$\therefore \quad \mathbf{W} = [\mathbf{A}^{-1}]^t\mathbf{S} = [\mathbf{A}^{-1}]^t\bar{\mathbf{k}}\boldsymbol{\alpha} = [\mathbf{A}^{-1}]^t\bar{\mathbf{k}}\mathbf{A}^{-1}\mathbf{u}$$

Hence

$$\mathbf{k} = [\mathbf{A}^{-1}]^t\bar{\mathbf{k}}\mathbf{A}^{-1}.$$

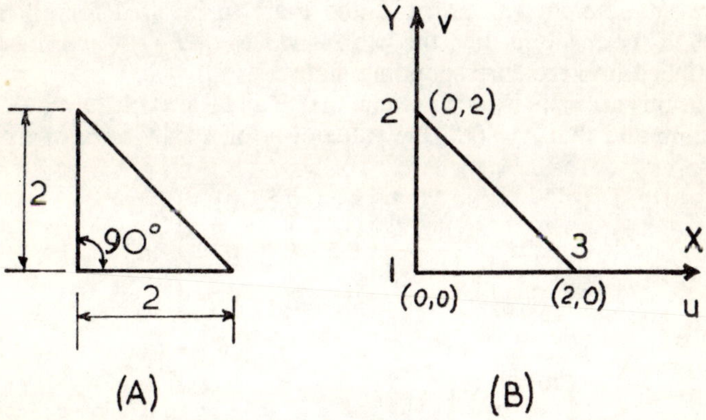

Fig. 7.3.

We are now in a position to determine the horizontal and vertical components of the loads applied at the nodal points, provided that we know the corresponding u and v components of displacement, by using the expression:

$$\mathbf{W} = \mathbf{k}\begin{bmatrix} u \\ \mathrm{v} \end{bmatrix}$$

Example

Let us determine the stiffness of the element shown in Figure 7.3A.

The first step is to express the element in local coordinates (Figure 7.3B). It has been established that $\bar{\mathbf{k}} = T \iint \mathbf{B}^t\mathbf{D}\mathbf{B}\, dx\, dy$. In a complete analysis the term T is carried through all the calculations to finally disappear when the stress values that we wish to find have been determined. Hence we may ignore this term from the start, by assuming it equal to unity, and take $\bar{\mathbf{k}}$ as being:

$$\bar{\mathbf{k}} = \iint \mathbf{B}^t\mathbf{D}\mathbf{B}\, dx\, dy$$

where

$$\mathbf{B} = \begin{bmatrix} 0 & 1 & 0 & 0 & 0 & 0 \\ 0 & 0 & 0 & 0 & 0 & 1 \\ 0 & 0 & 1 & 0 & 1 & 0 \end{bmatrix} \quad \text{and } \mathbf{D} = \frac{E}{1 - v^2}\begin{bmatrix} 1 & v & 0 \\ v & 1 & 0 \\ 0 & 0 & \dfrac{1 - v}{2} \end{bmatrix}$$

If E and v are taken as constant over the continuum then the term $E/(1 - v^2)$ need not be evaluated until the end of the analysis. If, however, different values for E and v are to be used for different parts of the medium then the various values of $E/(1 - v^2)$ must be multiplied into the corresponding matrices.

For this example we will assume that E and v are constant over the medium and that $v = 0{\cdot}5$ (the value of v for a fully saturated soil). i.e.

$$\mathbf{D} = \frac{E}{1 - v^2}\begin{bmatrix} 1 & 0{\cdot}5 & 0 \\ 0{\cdot}5 & 1 & 0 \\ 0 & 0 & 0{\cdot}25 \end{bmatrix}$$

Hence

$$\bar{\mathbf{k}} = \frac{E}{1 - v^2}\iint \begin{bmatrix} 0 & 0 & 0 \\ 1 & 0 & 0 \\ 0 & 0 & 1 \\ 0 & 0 & 0 \\ 0 & 0 & 1 \\ 0 & 1 & 0 \end{bmatrix} \begin{bmatrix} 1 & 0{\cdot}5 & 0 \\ 0{\cdot}5 & 1 & 0 \\ 0 & 0 & 0{\cdot}25 \end{bmatrix}$$

$$\times \begin{bmatrix} 0 & 1 & 0 & 0 & 0 & 0 \\ 0 & 0 & 0 & 0 & 0 & 1 \\ 0 & 0 & 1 & 0 & 1 & 0 \end{bmatrix} dx\, dy$$

i.e.

$$\bar{\mathbf{k}} = \frac{E}{1 - v^2} \iint \begin{bmatrix} 0 & 0 & 0 & 0 & 0 & 0 \\ 0 & 1 & 0 & 0 & 0 & 0{\cdot}5 \\ 0 & 0 & 0{\cdot}25 & 0 & 0{\cdot}25 & 0 \\ 0 & 0 & 0 & 0 & 0 & 0 \\ 0 & 0 & 0{\cdot}25 & 0 & 0{\cdot}25 & 0 \\ 0 & 0{\cdot}5 & 0 & 0 & 0 & 1 \end{bmatrix} \mathrm{d}x\,\mathrm{d}y$$

THE DOUBLE INTEGRATION OPERATION

The mathematical symbols

$$\int_{y_1}^{y_2} \int_{x_1}^{x_2} f(x, y)\,\mathrm{d}x\,\mathrm{d}y$$

mean that the function (x, y) is to be integrated with respect to x between the limits x_1 to x_2 and the result is then to be integrated with respect to y, between the limits y_1 to y_2. To carry out this double integration on the matrix $\mathbf{B}^t\mathbf{D}\mathbf{B}$ the operation must be conducted term by term.

If we consider the general term a_{ij} the expression can be written as

$$\bar{k}_{ij} = \frac{E}{1 - v^2} \iint a_{ij}\,\mathrm{d}x\,\mathrm{d}y$$

and, as a_{ij} is a term independent of x and y, may be written as:

$$\bar{k}_{ij} = \frac{E}{1 - v^2} a_{ij} \iint \mathrm{d}x\,\mathrm{d}y = C \iint \mathrm{d}x\,\mathrm{d}y$$

where C is a constant. Hence

$$\bar{k}_{ij} = C \int (\int \mathrm{d}x)\,\mathrm{d}y$$

the integral in the bracket being evolved first.

$$= C \int x\,\mathrm{d}y$$

The relationship between x and y within the particular element being considered is given by

$$y = y_2 - \frac{y_2}{x_3} x$$

or

$$x = \frac{x_3}{y_2}(y_2 - y) = (2 - y)$$

when the values of x_3 and y_2 are inserted.

Hence

$$\bar{\mathbf{k}} = C\int_0^{y_2} (2 - y)\,dy$$

inserting the limits

$$= C\left[2y - \frac{y^2}{2}\right]_0^2$$

$$= C(4 - 2) = 2C$$

Hence, to carry out the integration every nonzero term of the matrix must be multiplied by $2 \times E/(1 - v^2)$.

It will be seen that 2 is the area of the element and, therefore, in order to carry out integrations of this type on a triangular element all that is necessary is to multiply each nonzero term of the $\mathbf{B}^t\mathbf{DB}$ matrix by the area of the element. The $E/(1 - v^2)$ term will be placed either in front of or within the matrix depending upon whether or not E and v have been assumed constant over the whole medium. Use of the area of the element in this fashion is only possible if the coordinates of the nodes of the element are constant. In some finite element work these coordinates are themselves functions of x and y as is the case with a rectangular element for example. In these cases a solution can only be obtained by carrying out the formalised integration procedure briefly discussed above.

Hence, in this problem:

$$\bar{\mathbf{k}} = \frac{E}{1 - v^2}\begin{bmatrix} 0 & 0 & 0 & 0 & 0 & 0 \\ 0 & 2 & 0 & 0 & 0 & 1 \\ 0 & 0 & 0{\cdot}5 & 0 & 0{\cdot}5 & 0 \\ 0 & 0 & 0 & 0 & 0 & 0 \\ 0 & 0 & 0{\cdot}5 & 0 & 0{\cdot}5 & 0 \\ 0 & 1 & 0 & 0 & 0 & 2 \end{bmatrix}$$

To obtain the coordinate elemental stiffness, $\mathbf{k}$, we employ the transformation $\mathbf{k} = [\mathbf{A}^{-1}]^t\bar{\mathbf{k}}\mathbf{A}^{-1}$. Now

$$\mathbf{A} = \begin{bmatrix} 1 & 0 & 0 & 0 & 0 & 0 \\ 1 & x_2 & y_2 & 0 & 0 & 0 \\ 1 & x_3 & y_3 & 0 & 0 & 0 \\ 0 & 0 & 0 & 1 & 0 & 0 \\ 0 & 0 & 0 & 1 & x_2 & y_2 \\ 0 & 0 & 0 & 1 & x_3 & y_3 \end{bmatrix} = \begin{bmatrix} 1 & 0 & 0 & 0 & 0 & 0 \\ 1 & 0 & 2 & 0 & 0 & 0 \\ 1 & 2 & 0 & 0 & 0 & 0 \\ 0 & 0 & 0 & 1 & 0 & 0 \\ 0 & 0 & 0 & 1 & 0 & 2 \\ 0 & 0 & 0 & 1 & 2 & 0 \end{bmatrix}$$

The inverse of **A** can be readily obtained by partitioning the matrix, as explained in Chapter 3.

$$\mathbf{A}^{-1} = \begin{bmatrix} 1 & 0 & 0 & 0 & 0 & 0 \\ -0{\cdot}5 & 0 & 0{\cdot}5 & 0 & 0 & 0 \\ -0{\cdot}5 & 0{\cdot}5 & 0 & 0 & 0 & 0 \\ 0 & 0 & 0 & 1 & 0 & 0 \\ 0 & 0 & 0 & -0{\cdot}5 & 0 & 0{\cdot}5 \\ 0 & 0 & 0 & -0{\cdot}5 & 0{\cdot}5 & 0 \end{bmatrix}$$

And

$$\mathbf{A}^{-1t} = \begin{bmatrix} 1 & -0{\cdot}5 & -0{\cdot}5 & 0 & 0 & 0 \\ 0 & 0 & 0{\cdot}5 & 0 & 0 & 0 \\ 0 & 0{\cdot}5 & 0 & 0 & 0 & 0 \\ 0 & 0 & 0 & 1 & -0{\cdot}5 & -0{\cdot}5 \\ 0 & 0 & 0 & 0 & 0 & 0{\cdot}5 \\ 0 & 0 & 0 & 0 & 0{\cdot}5 & 0 \end{bmatrix}$$

Hence

$$\mathbf{k} = \mathbf{A}^{-1t}\bar{\mathbf{k}}\mathbf{A}^{-1}$$

$$= \frac{E}{1 - v^2}\begin{bmatrix} 0{\cdot}625 & -0{\cdot}125 & -0{\cdot}5 & 0{\cdot}375 & -0{\cdot}25 & -0{\cdot}125 \\ -0{\cdot}125 & 0{\cdot}125 & 0 & -0{\cdot}125 & 0 & 0{\cdot}125 \\ -0{\cdot}5 & 0 & 0{\cdot}5 & -0{\cdot}25 & 0{\cdot}25 & -0{\cdot}125 \\ 0{\cdot}375 & -0{\cdot}125 & -0{\cdot}25 & 0{\cdot}625 & -0{\cdot}5 & -0{\cdot}125 \\ -0{\cdot}25 & 0 & 0{\cdot}25 & -0{\cdot}5 & 0{\cdot}5 & 0 \\ -0{\cdot}125 & 0{\cdot}125 & 0 & -0{\cdot}125 & 0 & 0{\cdot}125 \end{bmatrix}$$

THE COMPLETE OR OVERALL STIFFNESS MATRIX, K

If we determine the overall stiffness matrix of the medium then we can write down an expression relating all the applied loads and reactions to the displacements at all the nodal points:

$$\mathbf{W} = \mathbf{K}\begin{bmatrix} u \\ \mathrm{v} \end{bmatrix}$$

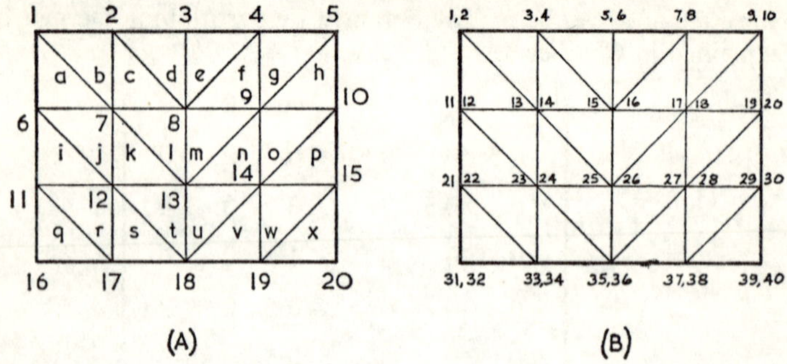

Fig. 7.4.

where **W** = the load matrix and

$$\begin{bmatrix} u \\ v \end{bmatrix}$$

is a vector listing all the nodal displacements.

One method to determine the elements of the overall stiffness matrix is to simply add the stiffnesses of all the elements adjacent to the nodal point considered. Other techniques are available but the method that will be described has the advantages of simplicity and that it uses less computer storage.

Let us assume that a particular medium has been split into 24 elements, *a*, *b*, *c*, . . ., *v*, *w*, *x*, for the purpose of analysis. (Figure 7.4). The nodal points have been numbered from 1 to 20 as shown.

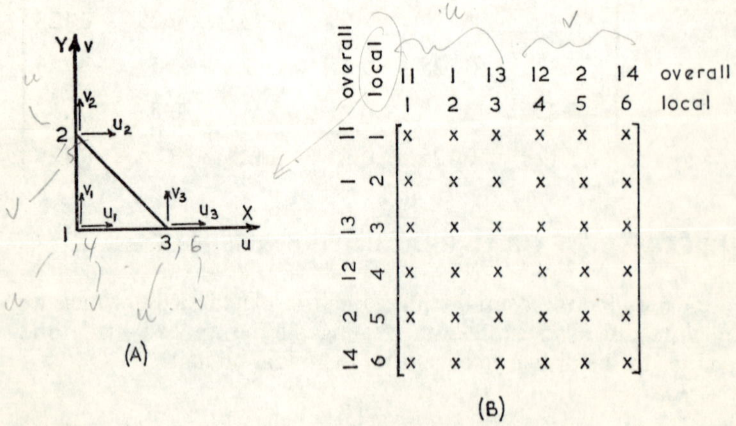

Fig. 7.5.

If at this stage we ignore any boundary conditions then every nodal point can be assumed to have two degrees of freedom, i.e. it is capable of displacement in both the horizontal and the vertical directions. At each nodal point let us assign a number to represent the two displacement components, say odd numbers to represent horizontal displacements and even numbers to represent vertical displacements. The numbering of these displacement components is shown in Figure 7.4B.

Consider the nodal point 1. There are two elements adjacent to this point, *a* and *b*. The two components of displacement are 1 and 2. Let us consider element *a*. The convention used to determine the stiffness, *k*, of this element is shown in Figure 7.5A.

Hence the local coordinate points 1, 2 and 3 of the element represent the actual nodal points 6, 1 and 7. In other words:

u_1 of element *a* represents the displacement component 11

u_2 of element *a* represents the displacement component 1

u_3 of element *a* represents the displacement component 13

v_1 of element *a* represents the displacement component 12

v_2 of element *a* represents the displacement component 2

v_3 of element *a* represents the displacement component 14

It is seen therefore that the first row of the elemental stiffness matrix of element *a* must be part of the eleventh row of the overall stiffness matrix. Similarly the first column of the elemental stiffness matrix must be part of the eleventh column of the overall stiffness matrix. The corresponding rows and columns are shown in Figure 7.5B.

For example the term $a_{1,1}$ of the elemental stiffness matrix is the term $a_{11,11}$ of the overall stiffness matrix. The term $a_{3,6}$ of the elemental stiffness matrix is the term $a_{13,14}$ of the overall stiffness matrix. If the convention to determine *k* for element *b* was as shown in Figure 7.6A then the relationship between the two sets of rows and columns will be as shown in Figure 7.6B.

The positions of all these terms in a portion of the overall stiffness matrix are shown in Figure 7.7. The terms of the stiffness matrix for element *a* are marked *x* and the terms for element *b* are marked 0. It is seen that there are several points where *x* and 0 appear together. At these points the relevant terms must be added together.

The process is continued, working from node to node. Each time a particular element is encountered then the terms of its elemental stiffness must be added into the overall stiffness matrix. For example at joint 2 there are three elements, *b*, *c* and *d*. Hence the element *b*,

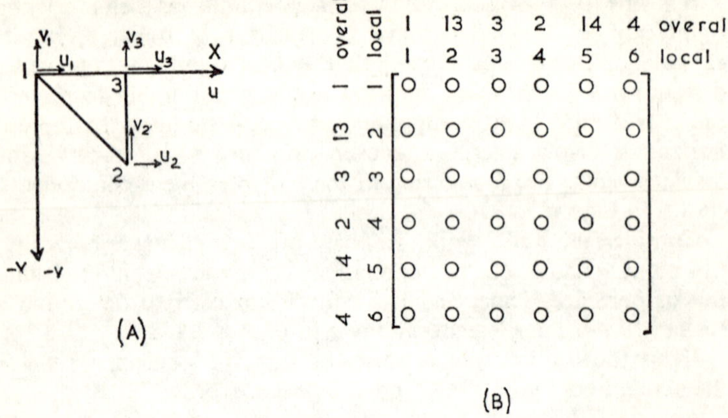

Fig. 7.6.

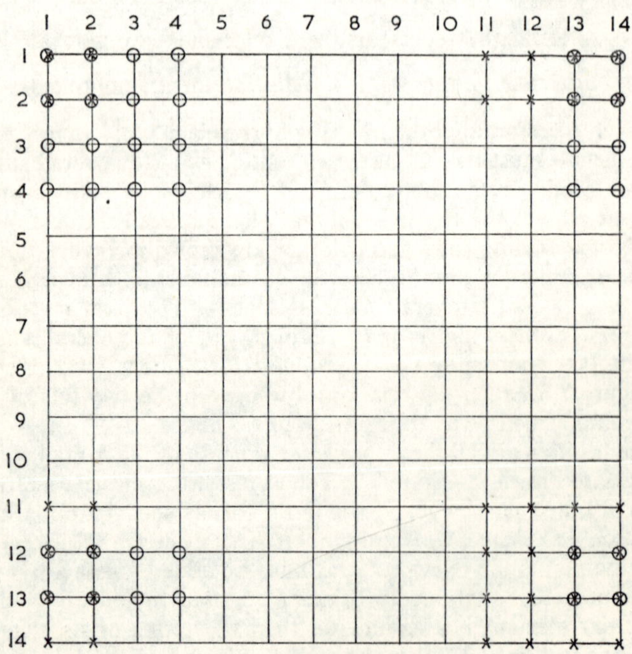

Fig. 7.7.

already included when considering joint 1, must be added in again. Following this procedure the 40 × 40 overall stiffness matrix will eventually be obtained. The form of this matrix is indicated in Figure 7.8 and it is seen that the matrix is heavily banded. The black dots represent the nonzero terms obtained from the elemental stiffnesses the rest of the points being zero terms. Some of the dots could of course represent zero values themselves if there were zero terms in some of the elemental stiffness matrices.

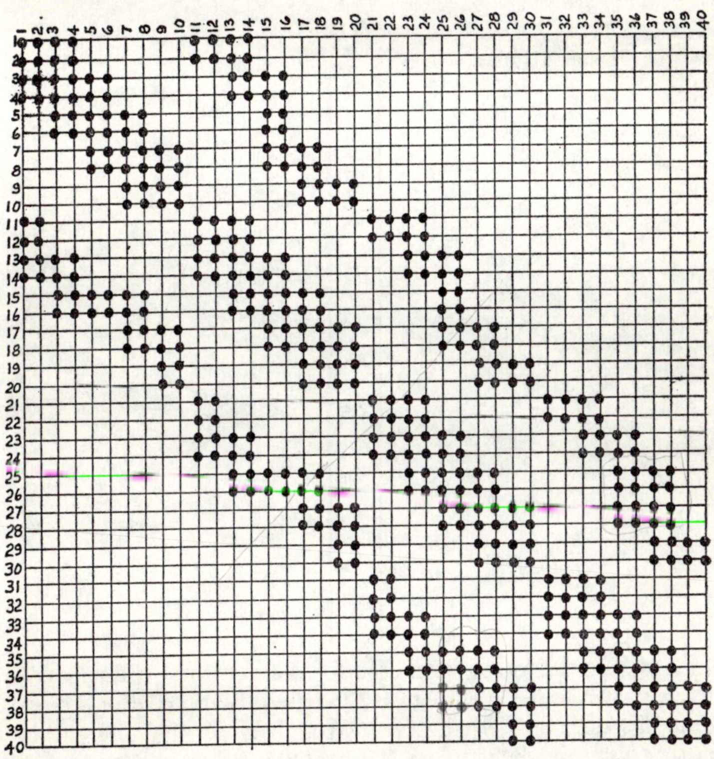

Fig. 7.8.

The matrix can be regarded as a band matrix containing two bands of zero elements. The zero elements that occur within the band indicate how some of the displacements are not connected to others. For example the displacements 5 and 6 are connected by element *d* to displacements 3 and 4 but are not connected to displacements 1 and 2.

Hence the terms $a_{3,5}$, $a_{5,3}$, $a_{4,5}$, $a_{5,4}$, $a_{3,6}$, $a_{6,3}$, $a_{4,6}$, $a_{6,4}$ must be nonzero terms, represented by black dots, whereas the terms $a_{1,5}$, $a_{5,1}$, $a_{2,5}$, $a_{5,2}$, $a_{1,6}$, $a_{6,1}$, $a_{2,6}$, $a_{6,2}$ must be zero terms.

It should be noted that, due to the format of the overall stiffness matrix, in the expression

$$\mathbf{W} = \mathbf{K}\begin{bmatrix} u \\ \mathrm{v} \end{bmatrix}$$

vector

$$\begin{bmatrix} u \\ \mathrm{v} \end{bmatrix}$$

must be in the form

$$\begin{bmatrix} u_1 \\ \mathrm{v}_1 \\ u_2 \\ \mathrm{v}_2 \\ u_3 \\ \mathrm{v}_3 \\ u_4 \\ \mathrm{v}_4 \\ \cdot \\ \cdot \\ \cdot \\ u_n \\ \mathrm{v}_n \end{bmatrix}$$

which is different to the form used for the elemental stiffnesses. The matrix shown in Figure 7.8 gives the relationship between the nodal point displacements in the form $\mathbf{W} = \mathbf{K}\boldsymbol{\Delta}$, where $\mathbf{W}$ is the load matrix (discussed later), $\mathbf{K}$ is the overall stiffness matrix, $\boldsymbol{\Delta}$ is the vector

$$\begin{bmatrix} u \\ \mathrm{v} \end{bmatrix}$$

representing displacements. If we obtain the inverse of $\mathbf{K}$ we have the relationship:

$$= \mathbf{K}^{-1}\mathbf{W} \qquad = \mathbf{FW}.$$

Where $\mathbf{F}$ is the overall flexibility matrix.

FURTHER NOTES ON THE ELEMENTAL STIFFNESS MATRIX

The reader may be concerned as to how to decide upon a local system of coordinates in order to evaluate **k** from $\bar{\mathbf{k}}$. There can be instances when a different stiffness matrix is evaluated with a different choice of coordinates. It must be appreciated that any arbitrary set of coordinates will lead to the same result. For example if we wished to evaluate the stiffness of the element shown in Figure 7.9A we might chose the local coordinate system shown in Figure 7.9B or the one shown in Figure 7.9C.

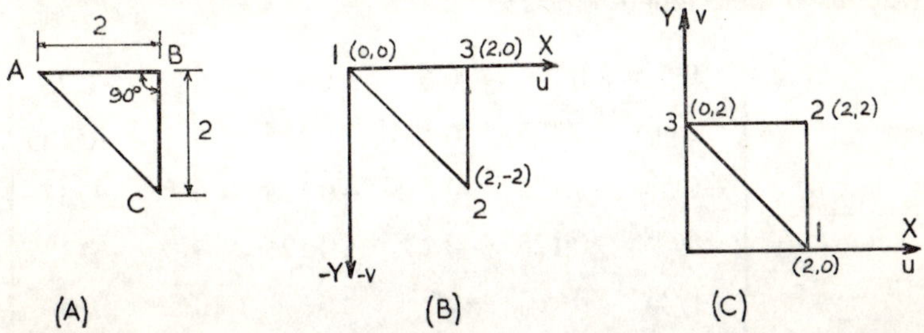

Fig. 7.9.

Assuming that $v = 0{\cdot}5$, $\bar{\mathbf{k}}$ will be as before:

$$\bar{\mathbf{k}} = \frac{E}{1-v^2}\begin{bmatrix} 0 & 0 & 0 & 0 & 0 & 0 \\ 0 & 2 & 0 & 0 & 0 & 1 \\ 0 & 0 & 0{\cdot}5 & 0 & 0{\cdot}5 & 0 \\ 0 & 0 & 0 & 0 & 0 & 0 \\ 0 & 0 & 0{\cdot}5 & 0 & 0{\cdot}5 & 0 \\ 0 & 1 & 0 & 0 & 0 & 2 \end{bmatrix}$$

Considering the coordinates shown in Figure 7.9B:

$$\mathbf{A} = \begin{bmatrix} 1 & 0 & 0 & 0 & 0 & 0 \\ 1 & 2 & -2 & 0 & 0 & 0 \\ 1 & 2 & 0 & 0 & 0 & 0 \\ 0 & 0 & 0 & 1 & 0 & 0 \\ 0 & 0 & 0 & 1 & 2 & -2 \\ 0 & 0 & 0 & 1 & 2 & 0 \end{bmatrix}$$

and

$$\mathbf{A}^{-1} = \begin{bmatrix} 1 & 0 & 0 & 0 & 0 & 0 \\ -0{\cdot}5 & 0 & 0{\cdot}5 & 0 & 0 & 0 \\ 0 & -0{\cdot}5 & 0{\cdot}5 & 0 & 0 & 0 \\ 0 & 0 & 0 & 1 & 0 & 0 \\ 0 & 0 & 0 & -0{\cdot}5 & 0 & 0{\cdot}5 \\ 0 & 0 & 0 & 0 & -0{\cdot}5 & 0{\cdot}5 \end{bmatrix}$$

leading to the evaluation of **k**:

$$\mathbf{k} = \frac{E}{1-v^2}\begin{bmatrix} 0{\cdot}5 & 0 & -0{\cdot}5 & 0 & 0{\cdot}25 & -0{\cdot}25 \\ 0 & 0{\cdot}125 & -0{\cdot}125 & 0{\cdot}125 & 0 & -0{\cdot}125 \\ -0{\cdot}5 & -0{\cdot}125 & 0{\cdot}625 & -0{\cdot}125 & -0{\cdot}25 & 0{\cdot}375 \\ 0 & 0{\cdot}125 & -0{\cdot}125 & 0{\cdot}125 & 0 & -0{\cdot}125 \\ 0{\cdot}25 & 0 & -0{\cdot}25 & 0 & 0{\cdot}5 & -0{\cdot}5 \\ -0{\cdot}25 & -0{\cdot}125 & 0{\cdot}375 & -0{\cdot}125 & -0{\cdot}5 & 0{\cdot}625 \end{bmatrix}$$

Considering the coordinates shown in Figure 7.9C:

$$\mathbf{A} = \begin{bmatrix} 1 & 2 & 0 & 0 & 0 & 0 \\ 1 & 2 & 2 & 0 & 0 & 0 \\ 1 & 0 & 2 & 0 & 0 & 0 \\ 0 & 0 & 0 & 1 & 2 & 0 \\ 0 & 0 & 0 & 1 & 2 & 2 \\ 0 & 0 & 0 & 1 & 0 & 2 \end{bmatrix}$$

and

$$\mathbf{A}^{-1} = \begin{bmatrix} 1 & -1 & 1 & 0 & 0 & 0 \\ 0 & 0{\cdot}5 & -0{\cdot}5 & 0 & 0 & 0 \\ -0{\cdot}5 & 0{\cdot}5 & 0 & 0 & 0 & 0 \\ 0 & 0 & 0 & 1 & -1 & 1 \\ 0 & 0 & 0 & 0 & 0{\cdot}5 & -0{\cdot}5 \\ 0 & 0 & 0 & -0{\cdot}5 & 0{\cdot}5 & 0 \end{bmatrix}$$

leading to the evaluation of **k**:

$$\mathbf{k} = \frac{E}{1-\nu^2}\begin{bmatrix} 0{\cdot}125 & -0{\cdot}125 & 0 & 0 & -0{\cdot}125 & 0{\cdot}125 \\ -0{\cdot}125 & 0{\cdot}625 & -0{\cdot}5 & -0{\cdot}25 & 0{\cdot}375 & -0{\cdot}125 \\ 0 & -0{\cdot}5 & 0{\cdot}5 & 0{\cdot}25 & -0{\cdot}25 & 0 \\ 0 & -0{\cdot}25 & 0{\cdot}25 & 0{\cdot}5 & -0{\cdot}5 & 0 \\ -0{\cdot}125 & 0{\cdot}375 & -0{\cdot}25 & -0{\cdot}5 & 0{\cdot}625 & -0{\cdot}125 \\ 0{\cdot}125 & -0{\cdot}125 & 0 & 0 & -0{\cdot}125 & 0{\cdot}125 \end{bmatrix}$$

Obviously the two matrices are different but they still express the same relationship, i.e.

$$\mathbf{W} = \mathbf{k}\begin{bmatrix} u \\ \mathrm{v} \end{bmatrix}$$

For example, let us evaluate the expression for a vertical load, V_B, acting at point B.

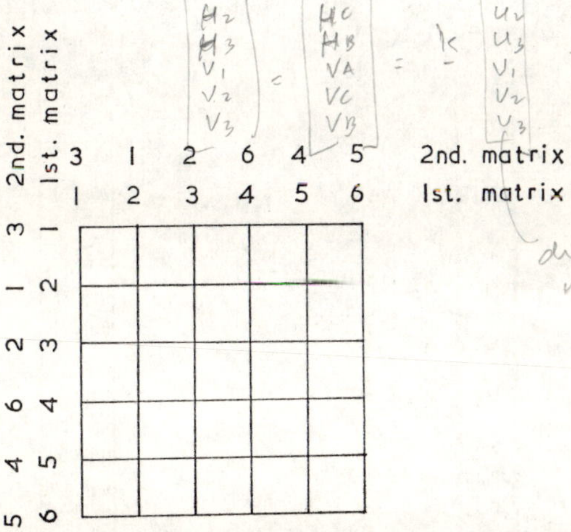

Fig. 7.10.

From the first matrix: (V_B is the 6th term of the load matrix)

$$V_B = -0{\cdot}25u_1 - 0{\cdot}125u_2 + 0{\cdot}375u_3 - 0{\cdot}125\mathrm{v}_1 - 0{\cdot}5\mathrm{v}_2 + 0{\cdot}625\mathrm{v}_3$$
$$-0{\cdot}25u_A - 0{\cdot}125u_C + 0{\cdot}375u_B - 0{\cdot}125\mathrm{v}_A - 0{\cdot}5\mathrm{v}_C + 0{\cdot}625\mathrm{v}_B$$

From the second matrix: (V_B is the 5th term of the load matrix)

$$V_B = -0{\cdot}125u_1 + 0{\cdot}375u_2 - 0{\cdot}25u_3 - 0{\cdot}5\mathrm{v}_1 + 0{\cdot}625\mathrm{v}_2 - 0{\cdot}125\mathrm{v}_3$$
$$= -0{\cdot}125u_C + 0{\cdot}375u_B - 0{\cdot}25u_A - 0{\cdot}5\mathrm{v}_C + 0{\cdot}625\mathrm{v}_B - 0{\cdot}125\mathrm{v}_A$$

The two expressions are identical. Indeed the second matrix can be evolved from the first by using the method for placing the terms of the elemental stiffnesses into the overall matrix, i.e. by exchanging corresponding rows and columns as indicated in Figure 7.10.

THE LOAD MATRIX

The stiffness matrix indicated in Figure 7.8 defines the structural properties of the medium to be analysed.

It is necessary to specify the loads to which this structure will be subjected. The loads are formed into a column vector each element of which corresponds to one of the displacement components u and v. Obviously the load components must be placed in the correct places.

Example

A plate is to be subjected to the load system shown in Figure (7.11). The plate has been split into six triangular elements and each nodal point is capable of both horizontal and vertical displacement.

The displacement matrix is therefore:

$$\begin{bmatrix} u_1 \\ v_1 \\ u_2 \\ v_2 \\ u_3 \\ v_3 \\ u_4 \\ v_4 \\ u_5 \\ v_5 \\ u_6 \\ v_6 \\ u_7 \\ v_7 \\ u_8 \\ v_8 \end{bmatrix}$$

The load matrix is therefore:

$$\begin{bmatrix} 0 \\ 0 \\ 0 \\ -4 \\ 0 \\ 0 \\ 0 \\ 3 \\ 2 \\ 0 \\ 0 \\ 0 \\ 0 \\ 0 \\ -2 \\ 0 \end{bmatrix}$$

Note that loads acting in the opposite direction to the directions taken as positive for displacements are negative, i.e. vertical loads acting downwards and horizontal loads acting to the left are negative.

It should be noted that only the applied loads, not the reactions, are listed in the load matrix. This may seem strange at first to an engineer used to determining the reactions as a first step in a structural analysis. However it is possible, in a framed structure, to work from the applied loads, through the frame, to eventually determine the reactions, it is this analogy that applies with finite elements. Once the elemental stresses have been determined the reactions can be found, if required.

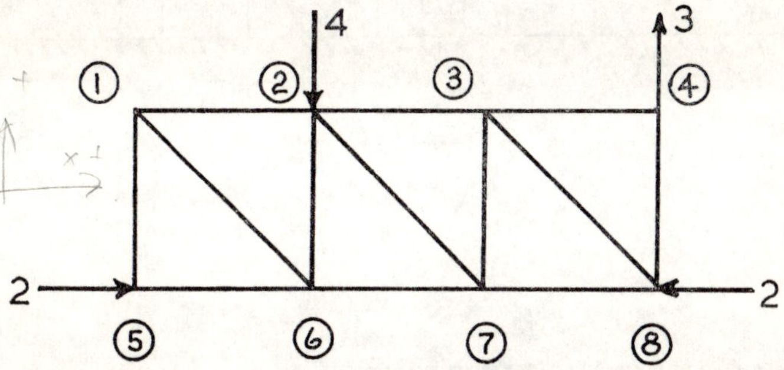

Fig. 7.11.

BOUNDARY CONDITIONS

In the solution of the type of problem discussed here it is necessary to assume boundaries in order to limit its size. The conditions along these boundaries can become difficult to determine and, with some methods of analysis, can involve a considerable amount of work. With the finite element method boundary conditions can often be handled by simply removing the corresponding rows and columns from the overall stiffness matrix.

Example 1

A finite layer of soil, overlying rigid rock and subjected to a continuous uniform foundation loading at its surface. This plane strain problem is illustrated in Figure 7.12. The foundation width is B and the lateral boundaries have been taken as CA and DB.

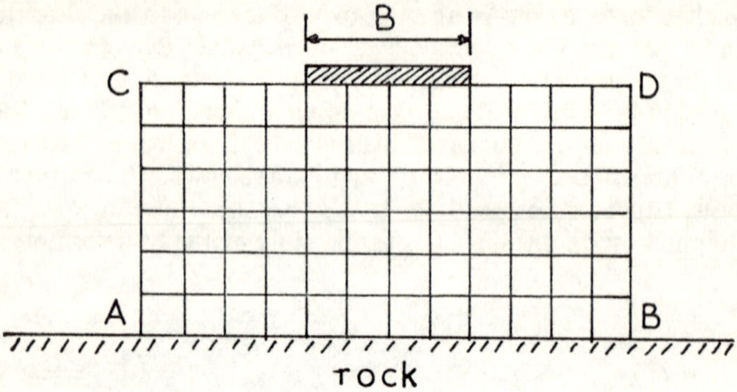

Fig. 7.12.

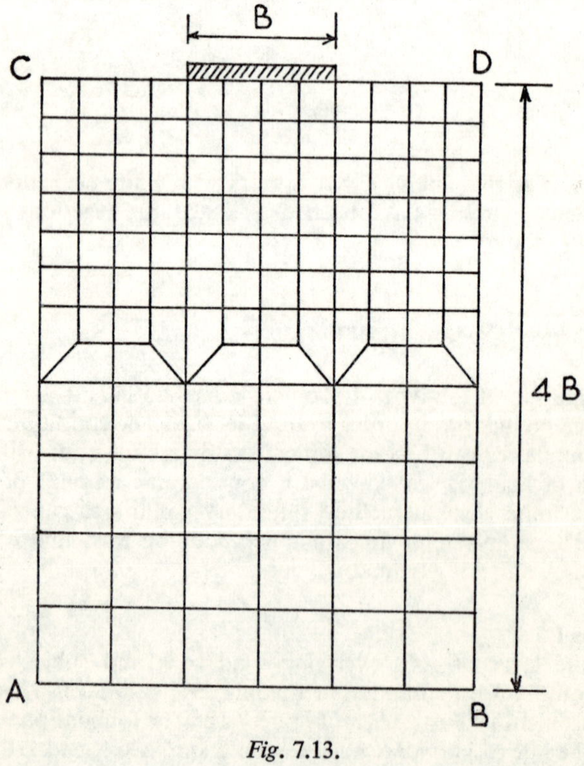

Fig. 7.13.

The interface between the rock and the soil, AB, is obviously a boundary at which there can be no vertical displacements of the soil but at which there could be horizontal movement. Hence, along the line AB, the vertical displacement components must be zero.

The lateral boundaries CA and DB must be taken far enough away from the foundation so that it may be assumed that the foundation loading has no effect on the points along CA and DB. Hence, along CA and DB the components of displacement, u and v, must be zero.

Example 2

A semi-infinite layer of soil subjected to a uniform foundation loading. The problem is illustrated in Figure 7.13. The lateral

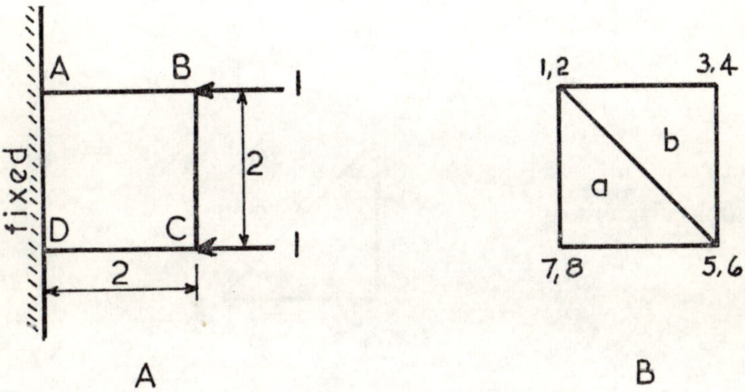

Fig. 7.14.

boundaries CA and DB will be taken as before but the horizontal boundary, AB, must be taken sufficiently below the foundation in order that it may be assumed that points along AB are unaffected by the loading.

Hence along AB the displacement components u and v are assumed to be zero. Note that the lower part of the grid is widened to allow a reduction in the calculations.

Example using finite elements

As an illustration of the finite element method let us consider the simple case of a thin square plate subjected in plane stress to two unit point loads applied at two corners, as shown in Figure 7.14A.

If we assume first that there are no boundary conditions then each corner is capable of two displacements. Let the horizontal displacements be 1, 3, 5 and 7 and let the vertical displacements be 2, 4, 6 and 8.

One form of finite element idealisation is shown in Figure 7.14B. This is the form that will be used here. If we assume that $v = 0{\cdot}5$ then the stiffness matrices of the elements a and b have already been evaluated in our earlier discussions.

$$\mathbf{k}_a = \frac{TE}{1 - v^2}\begin{bmatrix} 0{\cdot}625 & -0{\cdot}125 & -0{\cdot}5 & 0{\cdot}375 & -0{\cdot}25 & -0{\cdot}125 \\ -0{\cdot}125 & 0{\cdot}125 & 0 & -0{\cdot}125 & 0 & 0{\cdot}125 \\ -0{\cdot}5 & 0 & 0{\cdot}5 & -0{\cdot}25 & 0{\cdot}25 & 0 \\ 0{\cdot}375 & -0{\cdot}125 & -0{\cdot}25 & 0{\cdot}625 & -0{\cdot}5 & -0{\cdot}125 \\ -0{\cdot}25 & 0 & 0{\cdot}25 & -0{\cdot}5 & 0{\cdot}5 & 0 \\ -0{\cdot}125 & 0{\cdot}125 & 0 & -0{\cdot}125 & 0 & 0{\cdot}125 \end{bmatrix}$$

with local coordinates:

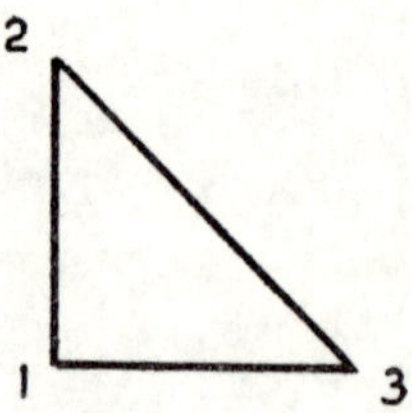

$$\mathbf{k}_b = \frac{TE}{1 - v^2}\begin{bmatrix} 0{\cdot}5 & 0 & -0{\cdot}5 & 0 & 0{\cdot}25 & -0{\cdot}25 \\ 0 & 0{\cdot}125 & -0{\cdot}125 & 0{\cdot}125 & 0 & -0{\cdot}125 \\ -0{\cdot}5 & -0{\cdot}125 & 0{\cdot}625 & -0{\cdot}125 & -0{\cdot}25 & 0{\cdot}375 \\ 0 & 0{\cdot}125 & -0{\cdot}125 & 0{\cdot}125 & 0 & -0{\cdot}125 \\ 0{\cdot}25 & 0 & -0{\cdot}25 & 0 & 0{\cdot}5 & -0{\cdot}5 \\ -0{\cdot}25 & -0{\cdot}125 & 0{\cdot}375 & -0{\cdot}125 & -0{\cdot}5 & 0{\cdot}625 \end{bmatrix}$$

with local coordinates:

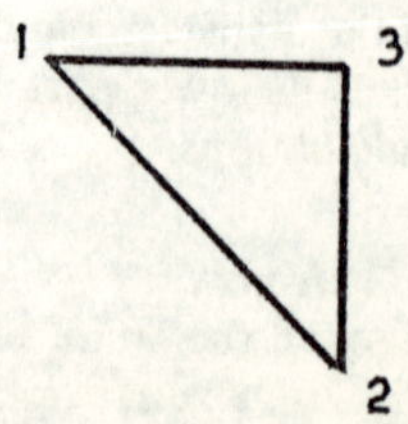

The overall stiffness matrix will be an 8 × 8 symmetrical matrix.

To insert the elements into the overall matrix the following convention must be used:

$\mathbf{k}_a$ (*row or col.*)	*overall* (*row or col.*)	$\mathbf{k}_b$ (*row or col.*)	*overall* (*row or col.*)
1	7	1	1
2	1	2	5
3	5	3	3
4	8	4	2
5	2	5	6
6	6	6	4

leading to the overall stiffness matrix:

$$\frac{TE}{1-v^2}\begin{bmatrix} 0{\cdot}625 & 0 & -0{\cdot}5 & -0{\cdot}25 & 0 & 0{\cdot}375 & -0{\cdot}125 & -0{\cdot}125 \\ 0 & 0{\cdot}625 & -0{\cdot}125 & -0{\cdot}125 & 0{\cdot}375 & 0 & -0{\cdot}25 & -0{\cdot}5 \\ -0{\cdot}5 & -0{\cdot}125 & 0{\cdot}625 & 0{\cdot}375 & -0{\cdot}125 & -0{\cdot}25 & 0 & 0 \\ -0{\cdot}25 & -0{\cdot}125 & 0{\cdot}375 & 0{\cdot}625 & -0{\cdot}125 & -0{\cdot}5 & 0 & 0 \\ 0 & 0{\cdot}375 & -0{\cdot}125 & -0{\cdot}125 & 0{\cdot}625 & 0 & -0{\cdot}5 & -0{\cdot}25 \\ 0{\cdot}375 & 0 & -0{\cdot}25 & -0{\cdot}5 & 0 & 0{\cdot}625 & -0{\cdot}125 & -0{\cdot}125 \\ -0{\cdot}125 & -0{\cdot}25 & 0 & 0 & -0{\cdot}5 & -0{\cdot}125 & 0{\cdot}625 & 0{\cdot}375 \\ -0{\cdot}125 & -0{\cdot}5 & 0 & 0 & -0{\cdot}25 & -0{\cdot}125 & 0{\cdot}375 & 0{\cdot}625 \end{bmatrix}$$

At this stage we can remove the columns and rows representing zero displacements which are known by the boundary conditions. As points A and D are fixed the displacements 1, 2, 7 and 8 must be zero. i.e. rows and columns numbers 1, 2, 7 and 8 are removed from the matrix to give the resultant overall stiffness matrix:

$$\frac{TE}{1-v^2}\begin{bmatrix} 0{\cdot}625 & 0{\cdot}375 & -0{\cdot}125 & -0{\cdot}25 \\ 0{\cdot}375 & 0{\cdot}625 & -0{\cdot}125 & -0{\cdot}5 \\ -0{\cdot}125 & -0{\cdot}125 & 0{\cdot}625 & 0 \\ -0{\cdot}25 & -0{\cdot}5 & 0 & 0{\cdot}625 \end{bmatrix} = \mathbf{K}$$

Hence we can write down the relationship:

$$\mathbf{W} = \mathbf{K\Delta},$$

where

$$\mathbf{W} = \begin{bmatrix} -1 \\ 0 \\ -1 \\ 0 \end{bmatrix} \quad \text{and} \quad \mathbf{\Delta} = \begin{bmatrix} u_3 \\ \mathrm{v}_4 \\ u_5 \\ \mathrm{v}_6 \end{bmatrix}$$

leading to the four simultaneous equations:

$$\begin{aligned} 0{\cdot}625u_3 + 0{\cdot}375\mathrm{v}_4 - 0{\cdot}125u_5 - 0{\cdot}25\mathrm{v}_6 &= -1 \\ 0{\cdot}375u_3 + 0{\cdot}625\mathrm{v}_4 - 0{\cdot}125u_5 - 0{\cdot}5\mathrm{v}_6 &= 0 \\ -0{\cdot}125u_3 - 0{\cdot}125\mathrm{v}_4 + 0{\cdot}625u_5 &= -1 \\ -0{\cdot}25u_3 - 0{\cdot}5\mathrm{v}_4 + 0{\cdot}625\mathrm{v}_6 &= 0 \end{aligned}$$

The solution of these equations gives:

$$\begin{aligned} u_3 &= -2{\cdot}710 \quad \text{direction } \leftarrow \\ \mathrm{v}_4 &= 1{\cdot}032 \quad \text{direction } \uparrow \\ u_5 &= -1{\cdot}935 \quad \text{direction } \leftarrow \\ \mathrm{v}_6 &= -0{\cdot}258 \quad \text{direction } \downarrow \end{aligned}$$

each value being divided by $TE/(1 - v^2)$.

An apparently more direct approach is to invert the stiffness matrix to give the relationship $\mathbf{FW} = \mathbf{\Delta}$, where

$$\mathbf{F} = \frac{1-v^2}{TE}\begin{bmatrix} 2{\cdot}581 & -1{\cdot}935 & 0{\cdot}129 & -0{\cdot}516 \\ -1{\cdot}935 & 6{\cdot}452 & 0{\cdot}903 & 4{\cdot}387 \\ 0{\cdot}129 & 0{\cdot}903 & 1{\cdot}806 & 0{\cdot}774 \\ -0{\cdot}516 & 4{\cdot}387 & 0{\cdot}774 & 4{\cdot}903 \end{bmatrix}$$

However, as will be seen later the flexibility method is not generally used.

TREATMENT OF BOUNDARY CONDITIONS

The reader may feel that, in the example, the work involved in finding the overall stiffness matrix, allowing for boundary conditions, would have been considerably simplified if the elements contained in the rows and columns finally eliminated had just not been calculated at all. This is perfectly correct. However, on a normal sized problem the number of rows and columns associated with the boundaries is very small compared to the overall number. It is usually simpler, in computer programming, to prepare the program so that the full matrix is first evolved before boundaries are considered.

VALIDITY OF RESULTS

It is obvious that the results obtained from the example are meaningless. In a perfect material, such as assumed, the horizontal displacements of points B and C must both be inwards and equal in magnitude.

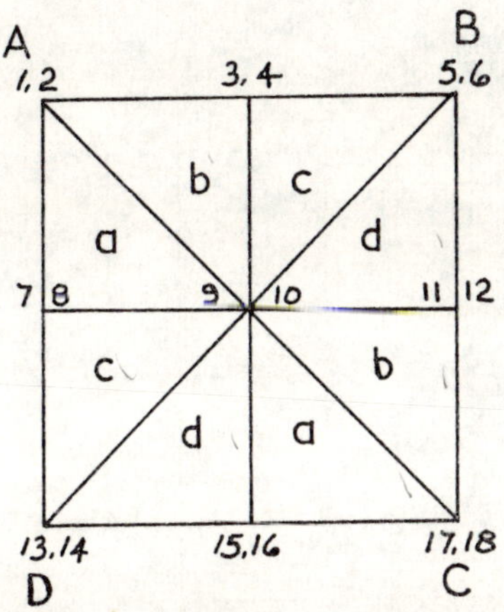

Fig. 7.15.

Similarly the vertical displacements of B and C must both be outwards and equal in magnitude.

The grid chosen was not sufficiently refined and was also not symmetrical. Symmetry is not important provided that enough elements are used although it is preferable to use a symmetrical pattern of elements where possible.

A further attempt at the solution of this problem would be to divide the plate into the twelve elements shown in Figure 7.15.

The stiffness matrices of the new elements a and b will have the same elements as the original matrices for a and b. This can be seen from the following argument:

In the determination of the original value of $\bar{\mathbf{k}}$ the integration procedure involved multiplying the matrix $\mathbf{B}^t\mathbf{DB}$ by the area of the element. The area was 2 but now, as the coordinate values have all been halved, the area is 1/2.

Hence
$$\text{new } \bar{\mathbf{k}} = \frac{\text{original } \bar{\mathbf{k}}}{4}$$

Also, as the coordinates have been halved;

$$\text{New matrix } \mathbf{A} = \tfrac{1}{2} \times \text{original matrix } \mathbf{A}$$

Hence
$$\text{new } \mathbf{A}^{-1} = 2 \times \text{original } \mathbf{A}^{-1}$$

Now
$$\text{new } \mathbf{k} = \mathbf{A}^{-1t}\bar{\mathbf{k}}\mathbf{A}^{-1} \text{ (new values)}$$
$$= 4\mathbf{A}^{-1t}\,\bar{\mathbf{k}}/4\;\mathbf{A}^{-1} \text{ (original values)}$$
$$= \text{original } \mathbf{k}$$

	3	4	5	6	9	10	11	12	15	16	17	18
3	1.25	0	-.5	.125	-.25	0						
4	0	1.25	.25	-.125	0	-1.0						
5	-.5	.25	.625	0	0	-.375	-.125	.125				
6	.125	-.125	0	.625	-.375	0	.25	-.5				
9	-.25	0	0	-.375	2.5	0	-1.0	0	-.25	0	0	.375
10	0	-1.0	-.375	0	0	2.5	0	-.25	0	-1.0	.375	0
11			-.125	.25	-1.0	0	1.25	0			-.125	-.25
12			.125	-.5	0	-.25	0	1.25			-.125	-.5
15					-.25	0			1.25	0	-.5	-.125
16					0	-1.0			0	1.25	-.25	-.125
17					0	.375	-.125	-.125	-.5	-.25	.625	0
18					.375	0	-.25	-.5	-.125	-.125	0	.625

Fig. 7.16.

If the reader is still not convinced he can always prepare the stiffness matrices for the new elements from the beginning.

The stiffness matrices for elements c and d evaluate to:

$$\mathbf{k}_c = \frac{TE}{1-v^2}\begin{bmatrix} 0{\cdot}625 & -0{\cdot}5 & -0{\cdot}125 & -0{\cdot}375 & 0{\cdot}125 & 0{\cdot}25 \\ -0{\cdot}5 & 0{\cdot}5 & 0 & 0{\cdot}25 & 0 & -0{\cdot}25 \\ -0{\cdot}125 & 0 & 0{\cdot}125 & 0{\cdot}125 & -0{\cdot}125 & 0 \\ -0{\cdot}375 & 0{\cdot}25 & 0{\cdot}125 & 0{\cdot}625 & -0{\cdot}125 & -0{\cdot}5 \\ 0{\cdot}125 & 0 & -0{\cdot}125 & -0{\cdot}125 & 0{\cdot}125 & 0 \\ 0{\cdot}25 & -0{\cdot}25 & 0 & -0{\cdot}5 & 0 & 0{\cdot}5 \end{bmatrix}$$

with local coordinates:

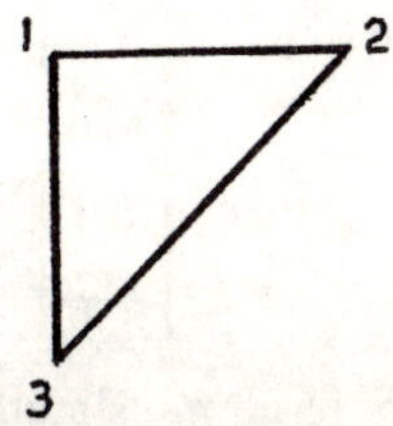

$$\mathbf{k}_d = \frac{TE}{1-v^2}\begin{bmatrix} 0{\cdot}5 & -0{\cdot}5 & 0 & 0 & 0{\cdot}25 & -0{\cdot}25 \\ -0{\cdot}5 & 0{\cdot}625 & -0{\cdot}125 & 0{\cdot}125 & -0{\cdot}375 & 0{\cdot}25 \\ 0 & -0{\cdot}125 & 0{\cdot}125 & -0{\cdot}125 & 0{\cdot}125 & 0 \\ 0 & 0{\cdot}125 & -0{\cdot}125 & 0{\cdot}125 & -0{\cdot}125 & 0 \\ 0{\cdot}25 & -0{\cdot}375 & 0{\cdot}125 & -0{\cdot}125 & 0{\cdot}625 & -0{\cdot}5 \\ -0{\cdot}25 & 0{\cdot}25 & 0 & 0 & -0{\cdot}5 & 0{\cdot}5 \end{bmatrix}$$

with local coordinates:

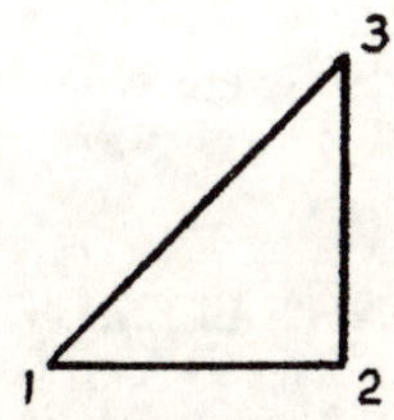

The final form of the overall stiffness matrix, allowing for the boundary conditions, is shown in Figure 7.16.

All unmarked elements in Figure 7.16 represent unconnected nodal points and have zero values, which have been omitted for clarity. With the same two horizontal unit loads applied at points B and C the load vector will be:

$$\begin{bmatrix} 0 \\ 0 \\ -1 \\ 0 \\ 0 \\ 0 \\ 0 \\ 0 \\ 0 \\ 0 \\ -1 \\ 0 \end{bmatrix}$$

The solution of the resulting simultaneous equations gives the following values:

$$\begin{aligned} u_3 &= -1{\cdot}627 & v_4 &= +0{\cdot}770 \\ u_5 &= -3{\cdot}524 & v_6 &= +0{\cdot}653 \\ u_9 &= -0{\cdot}758 & v_{10} &= 0{\cdot}0 \\ u_{11} &= -1{\cdot}573 & v_{12} &= 0{\cdot}0 \\ u_{15} &= -1{\cdot}627 & v_{16} &= -0{\cdot}770 \\ u_{17} &= -3{\cdot}524 & v_{18} &= -0{\cdot}653 \end{aligned}$$

It is seen that the results are symmetrical, as they should be, but there is no way of telling if they are as accurate as they might be.

FORM OF THE FINITE ELEMENT GRID

At the point of application of a concentrated load there is a region of high stress concentration. Such areas can also exist around holes, at sudden changes in stiffness, etc. In order to pick up the rapid changes in stress values in these regions it is necessary to include more elements than in a region further removed. As an example the stress distribution within the plate of the previous example would be shown up

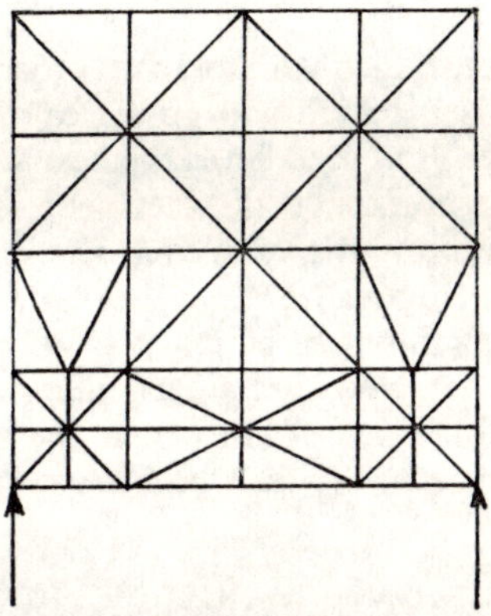

Fig. 7.17.

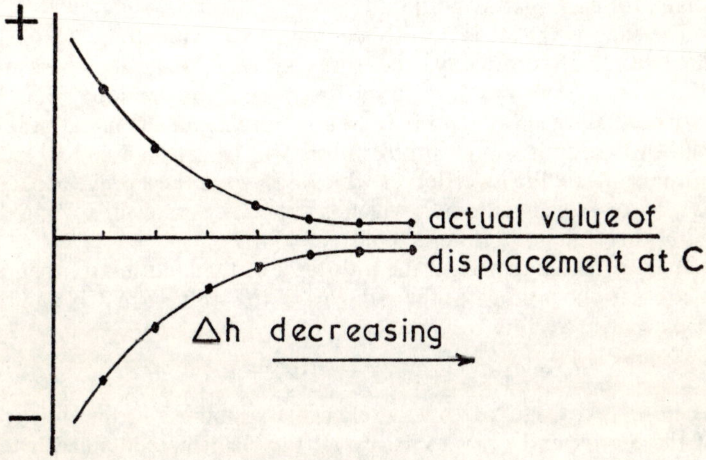

Fig. 7.18.

more sensibly if a grid similar to the one shown in Figure 7.17 was adopted. The diagram is purely illustrative. Obviously in such a problem a greater refinement of grid will be required.

ACCURACY OF RESULTS

It is possible to introduce an iterative process into a computer program so that the machine will go on calculating values with a diminishing grid size until it achieves the best accuracy. For example if the values of the horizontal displacement of the point C of the previous example were plotted for each size of grid interval, Δh, a convergence is seen (Figure 7.18).

The computer therefore can be made to continue subdividing the grid until the value obtained after a particular iteration is the same as the previous value obtained. At this stage the machine has reached the limit of accuracy and can then be made to print out the results.

SOLUTION OF FINITE ELEMENT SIMULTANEOUS EQUATIONS

At first glance the use of the flexibility matrix would appear to be the simplest method. The approach is more direct and, once the matrix has been established, it can be applied to any given load system to give values of displacements directly whereas, when using the stiffness matrix, a different set of simultaneous equations must be solved for each load condition.

The snag is that it is a fairly complicated matter to evaluate the flexibility matrix directly, the simplest way being to invert the stiffness matrix as was done in the first part of the previous example.

A normal sized problem may well involve some 600 nodal points. With a two dimensional problem this will lead to a 1,200 × 1,200 stiffness matrix the inversion of which is beyond the storage capacity of a normal sized computer, which is perfectly capable of solving the 1,200 corresponding simultaneous equations.

If we consider that R_n is the n-th term of the load matrix then R_n is equal to the product of the n-th row of the stiffness matrix and the displacement vector:

$$R_n = a_{n1}u_1 + a_{n2}\mathrm{v}_2 + a_{n3}u_3 + a_{n4}\mathrm{v}_4 + \dots$$

where a_{n1}, a_{n2}, etc. are all the elements contained in the n-th row of the matrix and u_1, v_2, etc. are all the elements contained in the displacement vector. In a large problem many of the elements in both the stiffness matrix and the load vector will be zero resulting in simultaneous equations that are relatively small.

There are several methods of solution available. The Gauss–Seidel method, discussed in Chapter 4, is easy to program, but the rate of convergence can be slow. A more direct method is by Gaussian elimination (Nat. Phys. Lab. 1961).

RECTANGULAR ELEMENTS IN PLANE STRESS

Plane stress and plane strain problems can be solved with the use of rectangular elements instead of triangles. Greater accuracy is obtained with a grid of rectangular elements compared to a grid with a similar number of triangular elements although more work is involved in the preparation of the elemental stiffnesses. Following this section is the complete procedure for determining the stiffness matrix of a rectangular element. The procedure appears lengthy but it should be remembered that much of the material consists of explanations and that, to a computer, the work is negligible.

STIFFNESS MATRIX OF A RECTANGULAR ELEMENT

A rectangular element is shown in Figure 7.19.

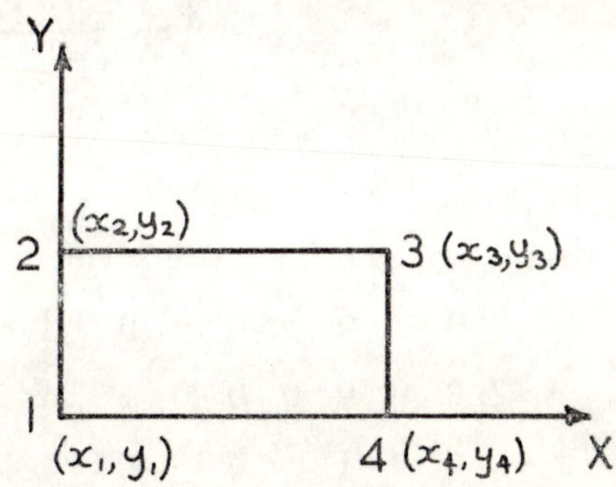

Fig. 7.19.

For the triangular element discussed previously we expressed u and v as linear functions of x and y, of the form $u(\mathrm{v}) = ax + by + c$. The coefficients of these equations were $\alpha_1 \ldots \alpha_6$ corresponding to the six degrees of freedom of the element.

In the case of a rectangular element, with four corners, there will

be eight degrees of freedom and Clough (1965) suggests that the displacement functions can be written as bilinear functions of x and y of the form $u(\mathrm{v}) = ax + bxy + cy + d$.
These expressions can be written as:

$$u = \alpha_1 + x\alpha_2 + y\alpha_3 + xy\alpha_4$$

$$\mathrm{v} = \alpha_5 + x\alpha_6 + y\alpha_7 + xy\alpha_8$$

or, in matrix form:

$$\begin{bmatrix} u \\ \mathrm{v} \end{bmatrix} = \begin{bmatrix} 1 & x & y & xy & 0 & 0 & 0 & 0 \\ 0 & 0 & 0 & 0 & 1 & x & y & xy \end{bmatrix} \begin{bmatrix} \alpha_1 \\ \cdot \\ \cdot \\ \cdot \\ \alpha_8 \end{bmatrix}$$

which is of the same form as for the triangular element, viz. $\mathbf{u} = \mathbf{A}\boldsymbol{\alpha}$.

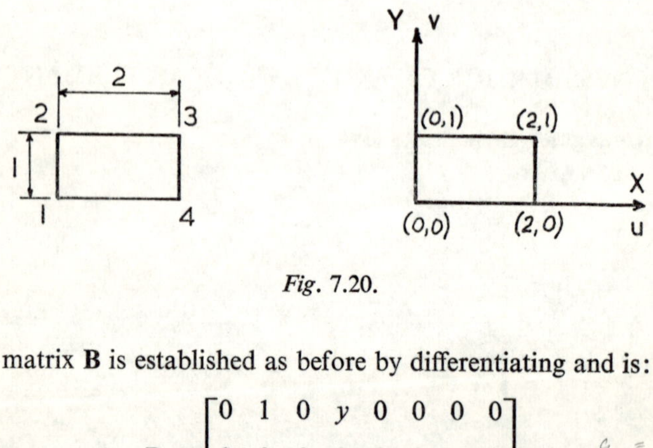

Fig. 7.20.

The matrix **B** is established as before by differentiating and is:

$$\mathbf{B} = \begin{bmatrix} 0 & 1 & 0 & y & 0 & 0 & 0 & 0 \\ 0 & 0 & 0 & 0 & 0 & 0 & 1 & x \\ 0 & 0 & 1 & x & 0 & 1 & 0 & y \end{bmatrix}$$

For elastic plane stress the matrix **D** will be as before:

$$\mathbf{D} = \frac{E}{1 - v^2} \begin{bmatrix} 1 & v & 0 \\ v & 1 & 0 \\ 0 & 0 & \dfrac{1 - v}{2} \end{bmatrix}$$

And the elemental stiffness,

$$\bar{\mathbf{k}} = T[\iint \mathbf{B}^t\mathbf{D}\mathbf{B}\,\mathrm{d}x\,\mathrm{d}y]$$

and

$$\mathbf{k} = [\mathbf{A}^{-1}]^t\bar{\mathbf{k}}\mathbf{A}^{-1}$$

Example

Suppose that a rectangular element has the dimensions shown in Figure 7.20A and that the local coordinate system adopted in order to determine its stiffness is as shown in Figure 7.20B.

From Figure 7.20B the matrix **A** can be quickly established from the general displacement functions:

$$\mathbf{A} = \begin{bmatrix} 1 & 0 & 0 & 0 & 0 & 0 & 0 & 0 \\ 1 & 0 & 1 & 0 & 0 & 0 & 0 & 0 \\ 1 & 2 & 1 & 2 & 0 & 0 & 0 & 0 \\ 1 & 2 & 0 & 0 & 0 & 0 & 0 & 0 \\ 0 & 0 & 0 & 0 & 1 & 0 & 0 & 0 \\ 0 & 0 & 0 & 0 & 1 & 0 & 1 & 0 \\ 0 & 0 & 0 & 0 & 1 & 2 & 1 & 2 \\ 0 & 0 & 0 & 0 & 1 & 2 & 0 & 0 \end{bmatrix}$$

$$\mathbf{B}^t\mathbf{D}\mathbf{B} = \frac{E}{1 - v^2}\begin{bmatrix} 0 & 0 & 0 \\ 1 & 0 & 0 \\ 0 & 0 & 1 \\ y & 0 & x \\ 0 & 0 & 0 \\ 0 & 0 & 1 \\ 0 & 1 & 0 \\ 0 & x & y \end{bmatrix}$$

$$\times \begin{bmatrix} 1 & 0{\cdot}5 & 0 \\ 0{\cdot}5 & 1 & 0 \\ 0 & 0 & 0{\cdot}25 \end{bmatrix} \begin{bmatrix} 0 & 1 & 0 & y & 0 & 0 & 0 & 0 \\ 0 & 0 & 0 & 0 & 0 & 0 & 1 & x \\ 0 & 0 & 1 & x & 0 & 1 & 0 & y \end{bmatrix}$$

(assuming $v = 0{\cdot}5$).

Hence

$$\mathbf{B}^{t}\mathbf{D}\mathbf{B} = \frac{E}{1-v^2}\begin{bmatrix} 0 & 0 & 0 & 0 & 0 & 0 & 0 & 0 \\ 0 & 1 & 0 & y & 0 & 0 & 0{\cdot}5 & 0{\cdot}5x \\ 0 & 0 & 0{\cdot}25 & 0{\cdot}25x & 0 & 0{\cdot}25 & 0 & 0{\cdot}25y \\ 0 & y & 0{\cdot}25x & (0{\cdot}25x^2 + y^2) & 0 & 0{\cdot}25x & 0{\cdot}5y & 0{\cdot}75xy \\ 0 & 0 & 0 & 0 & 0 & 0 & 0 & 0 \\ 0 & 0 & 0{\cdot}25 & 0{\cdot}25x & 0 & 0{\cdot}25 & 0 & 0{\cdot}25y \\ 0 & 0{\cdot}5 & 0 & 0{\cdot}5y & 0 & 0 & 1 & x \\ 0 & 0{\cdot}5x & 0{\cdot}25y & 0{\cdot}75xy & 0 & 0{\cdot}25y & x & (x^2 + 0{\cdot}25y^2) \end{bmatrix}$$

To obtain $\bar{\mathbf{k}}$ we must integrate the above matrix over the area of the element. Unlike the previous triangular element this matrix contains not only constant terms but also terms containing x, y, xy and $x^2 + y^2$. Let us consider the integration of each of these terms.

(i) Constant terms

Let the general constant term be $aE/(1 - v^2)$ then this term contributes $\bar{k}_{ij}$ to the elemental stiffness where

$$\bar{k}_{ij} = \frac{aE}{1 - v^2} \int_0^{y_2} \left[\int_0^{x_4} \mathrm{d}x \right] \mathrm{d}y$$

$$= \frac{aE}{1 - v^2} \int_0^1 \left[\int_0^2 \mathrm{d}x \right] \mathrm{d}y$$

$$= \frac{aE}{1 - v^2} \int_0^1 2\,\mathrm{d}y = \frac{2aE}{1 - v^2}$$

Note that as y is a constant value there is no relationship between x and y and the limits of the x integration must therefore be stated. As can be seen from the above, for a constant term, the integration procedure consists merely of multiplying the term by the area of the element, as would be expected by comparison with the triangular element.

(ii) Terms containing x

Let a general term be $bxE/(1 - v^2)$, then

$$\bar{k}_{ij} = \frac{bE}{1 - v^2} \int_0^1 \left[\int_0^2 x\,\mathrm{d}x \right] \mathrm{d}y \qquad \text{(a)}$$

$$= \frac{bE}{1 - v^2} \int_0^1 \left[\frac{x^2}{2} \right]_0^2 \mathrm{d}y$$

$$= \frac{2bE}{1 - v^2}$$

It will be noted that the integral part of Equation (a) represents the first moment of area of the element, about the y axis. Hence the term $bE/(1 - v^2)$ is simply multiplied by the first moment of area of the element about the y axis, in this case 2.

(iii) Terms containing y

Let the general term be $ayE/(1 - v^2)$.
Using a similar technique to the one described for the x term it can be shown that the integration procedure consists of multiplying the

term $ayE/(1 - v^2)$ by the first moment of area of the element about the x axis, in this case 1.

(iv) Terms containing xy

Let a general term be $bxyE/(1 - v^2)$, then

$$\bar{k}_{ij} = \frac{bE}{1 - v^2}\int_0^1 \left[\int_0^2 xy \, \mathrm{d}x\right] \mathrm{d}y \tag{b}$$

$$= \frac{bE}{1 - v^2}\int_0^1 \left[\frac{x^2}{2} y\right]_0^2 \mathrm{d}y$$

$$= \frac{bE}{1 - v^2}\int_0^1 2y \, \mathrm{d}y$$

$$= \frac{bE}{1 - v^2}\left[\frac{2y^2}{2}\right]_0^1$$

$$= \frac{bE}{1 - v^2}$$

It will be noted that the integral part of Equation (b) is the product of the moments of area, xy dx dy, or (xy dA), in this case 1.

(iv) Terms containing x^2 and y^2

Let a general term be $mx^2 + ny^2$

$$\bar{k}_{ij} = \frac{E}{1 - v^2}\int_0^1 \left[\int_0^2 (mx^2 + ny^2) \, \mathrm{d}x\right] \mathrm{d}y$$

$$= \frac{E}{1 - v^2}\int_0^1 \left[\frac{mx^3}{3} + ny^2x\right]_0^2 \mathrm{d}y$$

$$= \frac{E}{1 - v^2}\int_0^1 \left[\frac{8m}{3} + 2ny^2\right] \mathrm{d}y$$

$$= \frac{E}{1 - v^2}\left[\frac{8my}{3} + \frac{2ny^3}{3}\right]_0^1$$

$$= \frac{E}{1 - v^2}\left[\frac{8m}{3} + \frac{2n}{3}\right]$$

Inserting the above values in the appropriate places of the $\mathbf{B}^t\mathbf{D}\mathbf{B}$

matrix we obtain $\bar{\mathbf{k}}$:

$$\frac{E}{1-v^2}\begin{bmatrix} 0 & 0 & 0 & 0 & 0 & 0 & 0 & 0 \\ 0 & 2 & 0 & 1 & 0 & 0 & 1 & 1 \\ 0 & 0 & 0{\cdot}5 & 0{\cdot}5 & 0 & 0{\cdot}5 & 0 & 0{\cdot}25 \\ 0 & 1 & 0{\cdot}5 & 4/3 & 0 & 0{\cdot}5 & 0{\cdot}5 & 0{\cdot}75 \\ 0 & 0 & 0 & 0 & 0 & 0 & 0 & 0 \\ 0 & 0 & 0{\cdot}5 & 0{\cdot}5 & 0 & 0{\cdot}5 & 0 & 0{\cdot}25 \\ 0 & 1 & 0 & 0{\cdot}5 & 0 & 0 & 2 & 2 \\ 0 & 1 & 0{\cdot}25 & 0{\cdot}75 & 0 & 0{\cdot}25 & 2 & 8{\cdot}5/3 \end{bmatrix}$$

Now

$$\mathbf{A}^{-1} = \begin{bmatrix} 1 & 0 & 0 & 0 & 0 & 0 & 0 & 0 \\ -0{\cdot}5 & 0 & 0 & 0{\cdot}5 & 0 & 0 & 0 & 0 \\ -1 & 1 & 0 & 0 & 0 & 0 & 0 & 0 \\ 0{\cdot}5 & -0{\cdot}5 & 0{\cdot}5 & -0{\cdot}5 & 0 & 0 & 0 & 0 \\ 0 & 0 & 0 & 0 & 1 & 0 & 0 & 0 \\ 0 & 0 & 0 & 0 & -0{\cdot}5 & 0 & 0 & 0{\cdot}5 \\ 0 & 0 & 0 & 0 & -1 & 1 & 0 & 0 \\ 0 & 0 & 0 & 0 & 0{\cdot}5 & -0{\cdot}5 & 0{\cdot}5 & -0{\cdot}5 \end{bmatrix}$$

And

$$\mathbf{A}^{-1t} = \begin{bmatrix} 1 & -0{\cdot}5 & -1 & 0{\cdot}5 & 0 & 0 & 0 & 0 \\ 0 & 0 & 1 & -0{\cdot}5 & 0 & 0 & 0 & 0 \\ 0 & 0 & 0 & 0{\cdot}5 & 0 & 0 & 0 & 0 \\ 0 & 0{\cdot}5 & 0 & -0{\cdot}5 & 0 & 0 & 0 & 0 \\ 0 & 0 & 0 & 0 & 1 & -0{\cdot}5 & -1 & 0{\cdot}5 \\ 0 & 0 & 0 & 0 & 0 & 0 & 1 & -0{\cdot}5 \\ 0 & 0 & 0 & 0 & 0 & 0 & 0 & 0{\cdot}5 \\ 0 & 0 & 0 & 0 & 0 & 0{\cdot}5 & 0 & -0{\cdot}5 \end{bmatrix}$$

Hence

$$\mathbf{k} = \mathbf{A}^{-1t}\bar{\mathbf{k}}\mathbf{A}^{-1}$$

$$= \frac{1}{48}\frac{E}{1 - v^2}\begin{bmatrix} 16 & -4 & -8 & -4 & 9 & -3 & -9 & 3 \\ -4 & 16 & -4 & -8 & 3 & -9 & -3 & 9 \\ -8 & -4 & 16 & -4 & -9 & 3 & 9 & -3 \\ -4 & -8 & -4 & 16 & -3 & 9 & 3 & -9 \\ 9 & 3 & -9 & -3 & 34 & -31 & -17 & 14 \\ -3 & -9 & 3 & 9 & -31 & 34 & 14 & -17 \\ -9 & -3 & 9 & 3 & -17 & -14 & 34 & -31 \\ 3 & 9 & -3 & -9 & 14 & -17 & -31 & 34 \end{bmatrix}$$

As before the matrix can be roughly checked from the fact that it must be symmetrical and positive definite.

The matrix operations associated with rectangular elements are identical to those associated with triangular elements, the only difference being that the elemental stiffnesses are now represented by 8 × 8 matrices instead of 6 × 6. Any form of quadrilateral element is handled in the same manner as described for a rectangular element.

PLANE STRAIN

All that has been discussed so far in connection with plane stress is equally relevant to problems in plane strain.
The difference is that the matrix **D** for plane strain is:

$$\mathbf{D} = \frac{E(1 - v)}{(1 + v)(1 - 2v)}\begin{bmatrix} 1 & \frac{v}{(1 - v)} & 0 \\ \frac{v}{(1 - v)} & 1 & 0 \\ 0 & 0 & \frac{(1 - 2v)}{2(1 - v)} \end{bmatrix}$$

As explained in Chapter 6, values of $v = 0{\cdot}5$ render the formula inapplicable and it is therefore necessary, for such cases, to obtain an approximate solution by assuming $v = 0{\cdot}45$ or some other figure close to 0·5.

The reason that v has been taken as 0·5 for the plane stress examples was simply an attempt to keep the arithmetical work as simple as possible, for illustrative purposes, and has no other significance.

EVALUATION OF ELEMENTAL STRESSES

Before this technique is described a word of warning is not out of place. The examples in this chapter are unrealistic as they have been made small enough to allow the various matrix operations to be carried out by hand. The method of finite elements is intended for use with computers and the number of nodal points employed in a solution are considerably greater than the examples illustrated here, as mentioned previously, the employment of 600 nodal points is not uncommon. Nevertheless, with a relatively coarse grid it is possible to obtain reasonable values of displacements.

These remarks regarding coarse grid accuracy do not apply to the calculation of stress values. Stress values are second order functions in that they are obtained by using the derived displacement values. Hence, as one would intuitively expect, a considerably more refined grid is necessary when stress values are required.

The reader will appreciate that it is not possible to give a numerical example of the determination of stress values that would be small enough to allow inclusion in this book. All that can be done is to describe the procedure, assuming that the displacements have been evaluated to the necessary degree of accuracy. It has been shown that the stress to strain relationship for a particular element is given by:

$$\boldsymbol{\sigma} = \mathbf{D}\boldsymbol{\varepsilon} \quad \text{and, as} \quad \boldsymbol{\varepsilon} = \mathbf{B}\boldsymbol{\alpha}$$

Stresses can be expressed in terms of displacements by the relationship

$$\boldsymbol{\sigma} = \mathbf{DB}\boldsymbol{\alpha}$$

$$\boldsymbol{\sigma} = \mathbf{DBA}^{-1} \begin{bmatrix} u \\ \mathrm{v} \end{bmatrix}$$

Hence, once the nodal displacements have been obtained to the required accuracy the stresses in each element can be obtained from the relationship:

$$\begin{bmatrix} \sigma_x \\ \sigma_y \\ \tau_{xy} \end{bmatrix} = [\mathbf{DBA}^{-1}] \begin{bmatrix} u_1 \\ u_2 \\ u_3 \\ \mathrm{v}_1 \\ \mathrm{v}_2 \\ \mathrm{v}_3 \end{bmatrix}$$

where u_1, u_2 and u_3 are the horizontal displacements of points 1, 2 and 3 and v_1, v_2 and v_3 are the vertical displacements of points 1, 2 and 3.

Example

In the plane stress example the element *a* had the following properties:

$$\mathbf{D} = \frac{E}{1-v^2}\begin{bmatrix} 1 & 0{\cdot}5 & 0 \\ 0{\cdot}5 & 1 & 0 \\ 0 & 0 & 0{\cdot}25 \end{bmatrix} \quad \mathbf{B} = \begin{bmatrix} 0 & 1 & 0 & 0 & 0 & 0 \\ 0 & 0 & 0 & 0 & 0 & 1 \\ 0 & 0 & 1 & 0 & 1 & 0 \end{bmatrix}$$

$$\mathbf{A}^{-1} = 2\begin{bmatrix} 1 & 0 & 0 & 0 & 0 & 0 \\ -0{\cdot}5 & 0 & 0{\cdot}5 & 0 & 0 & 0 \\ -0{\cdot}5 & 0{\cdot}5 & 0 & 0 & 0 & 0 \\ 0 & 0 & 0 & 1 & 0 & 0 \\ 0 & 0 & 0 & -0{\cdot}5 & 0 & 0{\cdot}5 \\ 0 & 0 & 0 & -0{\cdot}5 & 0{\cdot}5 & 0 \end{bmatrix}$$

Hence

$$\mathbf{DBA}^{-1} = \frac{E}{1-v^2}\begin{bmatrix} -1 & 0 & 1 & -0{\cdot}5 & 0{\cdot}5 & 0 \\ -0{\cdot}5 & 0 & 0{\cdot}5 & -1 & 1 & 0 \\ -0{\cdot}25 & 0{\cdot}25 & 0 & -0{\cdot}25 & 0 & 0{\cdot}25 \end{bmatrix}$$

Considering the *a* element with nodal displacements 9, 10, 15, 16, 17 and 18 (Figure 7.15):

The displacements were found to be:

$$u_9 = -0{\cdot}758 \quad \mathrm{v}_{10} = 0{\cdot}0$$
$$u_{15} = -1{\cdot}627 \quad \mathrm{v}_{16} = -0{\cdot}77$$
$$u_{17} = -3{\cdot}524 \quad \mathrm{v}_{18} = -0{\cdot}653$$

each term being divided by $E/(1-v^2)$.

From the local coordinate system for this element, page 160, the corresponding displacements are:

point 1 – u_{15} and v_{16}

point 2 – u_9 and v_{10}

point 3 – u_{17} and v_{18}

Hence the expression becomes:

$$\begin{matrix}\sigma_x \\ \sigma_y \\ \tau_{xy}\end{matrix} = \frac{E}{1-v^2}\begin{bmatrix} -1 & 0 & 1 & -0{\cdot}5 & 0{\cdot}5 & 0 \\ -0{\cdot}5 & 0 & 0{\cdot}5 & -1 & 1 & 0 \\ -0{\cdot}25 & 0{\cdot}25 & 0 & -0{\cdot}25 & 0 & 0{\cdot}25 \end{bmatrix} \frac{1-v^2}{E}\begin{bmatrix} -1{\cdot}627 \\ -0{\cdot}758 \\ -3{\cdot}524 \\ -0{\cdot}77 \\ 0{\cdot}0 \\ -0{\cdot}653 \end{bmatrix}$$

leading to

$$\sigma_x = -1{\cdot}512 \quad \text{(compressive)}$$

See note 'Sign convention'

$$\sigma_y = -0{\cdot}179 \quad \text{(compressive)}$$

$$\tau_{xy} = +0{\cdot}247$$

As was stated at the beginning of this section the above values are meaningless. The example has been given to simply illustrate the method. In the solution of a practical problem a further iteration, with a more refined grid would now be carried out and the values compared with the previous set. This procedure has been illustrated in Figure 7.18. Instead of deflections, stress values across a section could be compared but these finer points cannot be examined at this introductory stage.

The three stress components thus obtained are considered to act at the centroid of the element. If required the principal stresses acting at this point can now be evaluated, using one of the methods described in Chapter 5.

By averaging out the stress on elements adjacent to a nodal point it is possible to arrive at the stress values at that point. This procedure has been described by Wilson (1963) but it is not often necessary to employ this technique.

UNITS

The units used in the solution of a problem must be consistent. For example if the grid dimensions were in metres, the applied loads in kilonewtons (kN) and E expressed in kN/m^2 then displacements would be in metres and stresses in kN/m^2.

SIGN CONVENTION

The object of this book is to try to describe current practice. In soil and rock mechanics, due to the preponderance of problems dealing with compressive stresses the convention has gradually been adopted that these stresses, and their resulting strains, are positive. The type of problems discussed in Chapters 5 and 6 are invariably expressed in this sign convention, hence its use in those chapters. It is perhaps a pity that this soils sign convention has evolved as it is the exact opposite of the one used in mechanical engineering and continuum mechanics. The mechanical sign convention is more logical in that an extension due to a tensile force, an increase in length, is regarded as causing a positive strain and that the force producing it is also positive.

In order to follow current literature the soils engineer and student is obliged to think in terms of two sign conventions:

(i) The normal soil mechanics convention, as described in Chapter 5, when determining a set of stress values from a known set and also in soil testing (i.e. the form of problems of the last two chapters).

(ii) The normal mechanical sign convention for stress analysis work which use techniques such as finite elements. The directions taken as positive for the displacements in the finite element method dictate that the signs of the final stresses will be in accordance with the mechanical convention, i.e. a normal stress with a negative sign is compressive, a normal stress with a positive sign is tensile and a shear stress with a positive sign represents a shear stress acting on the face of an element furthest from the origin and in a direction away from the origin.

8

Further Applications of the Finite Element Method

TREATMENT OF NONLINEAR STRESS–STRAIN RELATIONSHIPS

IN the last chapter the type of plane stress and strain problems discussed were linear in which, throughout the continuum, the relationship between stress and strain was defined by two absolute constants, E and v, which allowed the use of a matrix $\mathbf{D}$ of constant value. Such a condition forms the basis of the elastic theory and depends upon two factors: (i) The stress–strain relationship is of the form illustrated in Figure 6.1B, leading to a constant E value, (ii) Stresses throughout the medium never exceed the yield stress value.

INCREMENTAL LOAD PROCEDURE

In most metals and rocks the stress–strain relationship is approximately linear and, within the range of linearity, E can be assumed constant. Poisson's ratio, v, can be assumed to have a value of the order of 0·25.

For problems involving such materials the continuum is divided into elements and an increment of the applied load is assumed to be applied. Stresses and displacements are evaluated and the strain energy per unit volume of each element is determined. These values are compared with the value of strain energy density at which it is assumed plastic yield will occur. The E value of any element in which this value is exceeded is then assumed to be some small fraction of the elastic E (say 1/500). The incremental procedure is continued until the full load has been applied. At this stage the state of stress in the continuum, including the regions of plastic yield, will be known.

ITERATIVE PROCEDURE

In soils ideal conditions do not apply and there is no truly linear relationship between stress and strain, being more of the form shown in Figure 6.1C. The relationship is a varying factor and the slope of the stress to strain plot is no longer equal to the constant value E. There is the additional complication that the stress–strain relationship is not only nonlinear but is also time dependent. This latter effect becomes important in consolidation work but will not be considered here. The material in this section will deal with immediate loading effects only and the stress–strain curve will be assumed as in Figure 6.1C, a typical result from a quick triaxial compression test.

The procedure adopted is to use a pseudo-constant to express E. This term is known as the secant modulus and is simply the value of the ratio between stress and strain for a particular value of strain. Depending upon the value of the strain chosen so the secant modulus, which we will denote as E, will vary.

In the finite element method it is assumed that, within each element, the value of E is constant. To obtain a solution the calculations are started by assigning an approximate value of E for each element. With these values the strains at the centre of each element are evaluated. New values of E, in closer correspondence to the derived strain values are now inserted into the calculations which are repeated to produce new strain values. This iterative procedure is continued until the final set of strain values are within a suitable tolerance of the previous set.

It was seen in Chapter 6 that the three plastic yield theories give a measure of the octahedral shear stress at yield. It is often convenient to use octahedral stress and strains as this automatically takes into account the effect of all the principal stresses and strains. Yield theories, however, do not give an indication of the magnitude of strain at which failure will occur.

The expressions for octahedral normal and shear stresses have been derived in Chapter 5:

$$\sigma_{oct} = (\sigma_1 + \sigma_2 + \sigma_3)/3$$

$$\tau_{oct} = \tfrac{1}{3}\sqrt{[(\sigma_1 - \sigma_2)^2 + (\sigma_2 - \sigma_3)^2 + (\sigma_3 - \sigma_1)^2]}$$

A similar relationship applies to strains. The octahedral strain equations have not been derived but they can be written down by replacing each σ value in the above expressions by its corresponding ε and by replacing τ_{oct} by $\tfrac{1}{2}\gamma_{oct}$. Hence:

$$\varepsilon_{oct} = (\varepsilon_1 + \varepsilon_2 + \varepsilon_3)/3 \quad (1)$$

$$\gamma_{oct} = \tfrac{2}{3}\sqrt{[(\varepsilon_1 - \varepsilon_2)^2 + (\varepsilon_2 - \varepsilon_3)^2 + (\varepsilon_3 - \varepsilon_1)^2]} \quad (2)$$

Where ε_{oct} = octahedral lineal strain and γ_{oct} = octahedral shear strain. ε_1, ε_2 and ε_3 are the principal strains corresponding to the principal stresses.

It can be seen that, in a perfectly isotropic elastic material, ε_{oct} is equal to one third of the volumetric strain.

If the reader compares the above Equations (1 and 2) with Equations (4) and (5) of Chapter 6 he will see that, for a perfectly elastic material,

$$\tau_{oct} = G\,\gamma_{oct}$$

$$\sigma_{oct} = 3K\,\varepsilon_{oct}$$

It is therefore a reasonable assumption, in many materials, that the octahedral lineal strain is a function of the change in volume and that the octahedral shear strain is a function of the change of shape. i.e.

$$\varepsilon_{oct} = f(\tau_{oct})$$

$$\gamma_{oct} = f(\sigma_{oct})$$

IMMEDIATE SETTLEMENT OF A CIRCULAR FOUNDATION ON SOFT CLAY

Example

A finite element solution to this problem has been discussed by Girijavallabhan and Reese (1968). The problem is one of axial symmetry, which is described later, but the treatment of nonlinearity is of interest here.

During the initial stages of loading a foundation on clay the soil tends to act in the undrained state. There is no time for significant volume changes, due to consolidation, and the settlement of the foundation may be assumed to be due to a change in shape of the soil beneath it. In such a case v can be taken as 0·5 and the relevant laboratory test is the undrained triaxial.

In the triaxial test $\sigma_2 = \sigma_3$. Hence

$$\tau_{oct} = \tfrac{1}{3}\sqrt{[(\sigma_1 - \sigma_3)^2 + (\sigma_3 - \sigma_1)^2]}$$

$$= \tfrac{1}{3}\sqrt{(2\sigma_1{}^2 - 4\sigma_1\sigma_3 + 2\sigma_3{}^2)}$$

$$= \frac{\sqrt{2}}{3}(\sigma_1 - \sigma_3)$$

Also in the undrained test, as there is no change in volume:

$$\varepsilon_1 + \varepsilon_2 + \varepsilon_3 = 0$$

And as $\varepsilon_2 = \varepsilon_3$;

$$\varepsilon_3 = -\varepsilon_1/2$$

Hence

$$\tfrac{1}{2}\gamma_{\text{oct}} = \tfrac{1}{3}\sqrt{[(\varepsilon_1 + \tfrac{1}{2}\varepsilon_1)^2 + (-\tfrac{1}{2}\varepsilon_1 - \varepsilon_1)^2]}$$

i.e.

$$\gamma_{\text{oct}} = \sqrt{(2)}\,\varepsilon_1$$

In Chapter 6 it was established that

$$E = 2G(1 + v)$$
$$\approx 3G \quad (\text{for } v = 0{\cdot}5)$$

The relationships applicable to this problem are therefore:

$$\tau_{\text{oct}} = \frac{\sqrt{2}}{3}(\sigma_1 - \sigma_3) \tag{3}$$

$$\gamma_{\text{oct}} = \sqrt{(2)}\,\varepsilon_1 \tag{4}$$

$$G = \tau_{\text{oct}}/\gamma_{\text{oct}} \tag{5}$$

$$E = 3G \tag{6}$$

A typical deviator stress to axial strain curve obtained from an undrained triaxial test on a soft clay sample is shown in Figure 8.1A. For a saturated clay the stress to strain curves are fairly

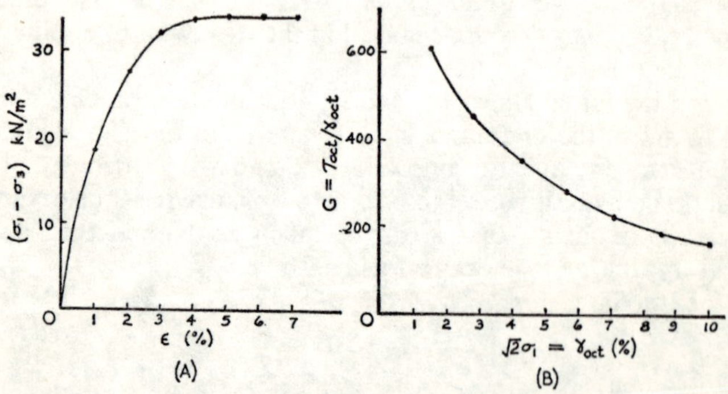

Fig. 8.1.

independent of the cell pressure used in the test and reasonably average values can be obtained.

From the experimental values the relationship between G and γ_{oct} can be established as set out in Table 8.1. For any particular value of γ_{oct}, E is approximately one third of the corresponding value of G. After each iteration the octahedral shear stresses at the centroid of

TABLE 8.1

% strain ε_1	$(\sigma_1 - \sigma_3)$ (kN/m^2)	$\frac{\sqrt{(2)}}{3}(\sigma_1 - \sigma_3) = \tau_{oct}$	$\sqrt{(2)}\varepsilon_1 = \gamma_{oct}$	$\tau_{oct}/\gamma_{oct} = G$ (kN/m^2)
1	18·5	8·71	·01414	616
2	27·5	12·95	·02828	458
3	32·2	15·17	·04242	358
4	33·8	15·92	·05656	282
5	34·0	16·02	·07070	225
6	34·0	16·02	·08484	188
7	34·0	16·02	·09898	162

each element are calculated and a corresponding new value for E inserted.

In this simple case v can be assumed constant. As the term $(1 - 2v)$ occurs in the **D** matrix of the axisymmetrical case a value slightly less than 0·5 must be used. Girijavallabhan and Reese took $v = 0{\cdot}48$.

STRESSES AND DISPLACEMENTS IN AN EARTH EMBANKMENT DURING CONSTRUCTION

Example

Constructional techniques dictate that most earth embankments and dams are built in definite increments so that loading and deformations occur simultaneously.

This problem has been discussed by Clough and Woodward (1967). The main assumption made was that the three dimensional problem could be reduced to a plane strain problem by considering the central cross section of the dam, normal to its axis. A further approximation was that the placed material was both isotropic and homogeneous. The authors allowed for the nonlinearity of the soil used in construction by the technique, briefly discussed below. The stress equations for plane strain have been derived in Chapter 6. Assuming that σ_1 and σ_3 are the principal stresses in the plane of strain:

$$\sigma_1 = \frac{E(1 - v)}{(1 + v)(1 - 2v)}\left[\varepsilon_1 + \frac{v}{1 - v}\varepsilon_3\right] \tag{7}$$

$$\sigma_3 = \frac{E(1 - v)}{(1 + v)(1 - 2v)}\left[\varepsilon_3 + \frac{v}{1 - v}\varepsilon_1\right] \tag{8}$$

The authors pointed out that the use of E and v (as in the last example) is not always the best approach and, for this type of problem, they suggest the use of two parameters, M_b and M_d such that

$$\frac{\sigma_1 + \sigma_3}{2} = M_b(\varepsilon_1 + \varepsilon_3)$$

$$\frac{\sigma_1 - \sigma_3}{2} = M_d(\varepsilon_1 - \varepsilon_3)$$

From Equations (7) and (8):

$$\begin{aligned}\sigma_1 + \sigma_3 &= \frac{E(1-v)}{(1+v)(1-2v)}\left[1 + \frac{v}{1-v}\right][\varepsilon_1 + \varepsilon_3] \\ &= \frac{E(1-v)}{(1+v)(1-2v)}\left[\frac{1-v+v}{1-v}\right][\varepsilon_1 + \varepsilon_3] \\ &= \frac{E}{(1+v)(1-2v)}[\varepsilon_1 + \varepsilon_3]\end{aligned}$$

Hence

$$\begin{aligned}\frac{\sigma_1 + \sigma_3}{2} &= \frac{E}{2(1+v)(1-2v)}[\varepsilon_1 + \varepsilon_3] \\ &= M_b(\varepsilon_1 + \varepsilon_3)\end{aligned}$$

where

$$M_b = \frac{E}{2(1+v)(1-2v)} \tag{9}$$

Similarly it can be shown that

$$M_d = E/(1+v) \tag{10}$$

i.e. the Rigidity Modulus, G. The elasticity matrix for plane strain can be written as:

$$\begin{bmatrix}\sigma_1 \\ \sigma_3 \\ \tau_{1,3}\end{bmatrix} = \frac{E}{(1+v)(1-2v)}\begin{bmatrix}(1-v) & v & 0 \\ v & (1-v) & 0 \\ 0 & 0 & \dfrac{1-2v}{2}\end{bmatrix}\begin{bmatrix}\varepsilon_1 \\ \varepsilon_3 \\ \gamma_{1,3}\end{bmatrix}$$

By introducing M_b and M_d the relationship becomes:

$$\begin{bmatrix}\sigma_1 \\ \sigma_3 \\ \tau_{1,3}\end{bmatrix} = \begin{bmatrix}(M_b + M_d) & (M_b - M_d) & 0 \\ (M_b - M_d) & (M_b + M_d) & 0 \\ 0 & 0 & M_d\end{bmatrix}\begin{bmatrix}\varepsilon_1 \\ \varepsilon_3 \\ \gamma_{1,3}\end{bmatrix}$$

To establish M_b and M_d a series of undrained triaxial tests are carried out on a series of samples of the fill material, compacted to the anticipated density, at different values of cell pressure. A typical set of results is illustrated in Figure 8.2 and it is seen that E, the secant modulus, varies with the cell pressure as well as the value of axial strain.

The procedure adopted by Clough was to first determine a value for the initial secant modulus by determining the slope of the curve for zero cell pressure and zero deviator stress and to assume an initial value for v (he used the value 0·44).

M_b and M_d were then determined from Equations (9) and (10). With this form of problem it is considered satisfactory if M_b is assumed to remain constant and to allow for the nonlinearity effects by varying M_d.

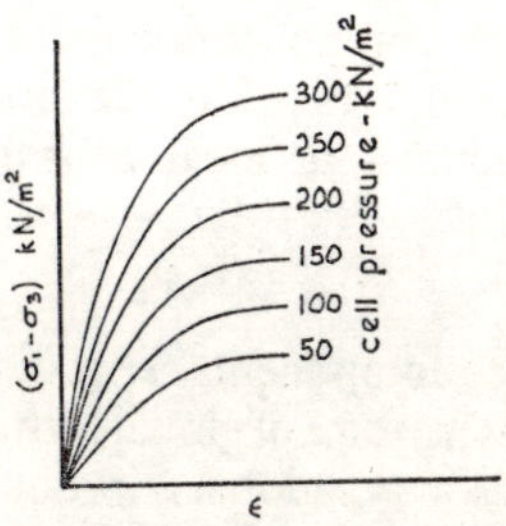

Fig. 8.2.

The first increment of construction is now assumed, suitable elements chosen, and the principal stresses evaluated at the centre of each element using the values of E, v, M_b and M_d so far determined. From σ_3 and $(\sigma_1 - \sigma_3)$ a new value of E is determined for each element from the test data represented in Figure 8.2. As M_b is assumed constant v can be obtained from Equation (9). Knowing E and v, M_d can now be evaluated from Equation (10).

New elemental stiffnesses are next evaluated and another construction increment assumed to be added. Principal stresses are now evaluated for all the elements and new values for v, E and M_d obtained for each element as before. The procedure is repeated until the last constructional lift has been allowed for.

In the above analysis no allowance is made for the settlement of the material supporting the dam. The finite element method can be used to allow for this effect, when it is considered necessary, and the technique has been described by Clough and Woodward.

LOCAL AND GLOBAL COORDINATES

Although no examples have been given there are occasions when, for convenience of computation, the stiffness of a particular element may have been defined in a local coordinate system which is not coincidental with the directions of the components of the applied forces and displacements of the continuum. These latter directions give the coordinates that apply to the whole problem and such coordinates are generally referred to as global coordinates. Obviously before the overall stiffness matrix can be assembled any displacements or nodal forces defined in a different local system must be transformed into the global system.

The problem therefore is that, knowing the relationship:

$$\mathbf{W}_L = \mathbf{k}_L \mathbf{u}_L$$

where $\mathbf{W}_L$ = nodal force components expressed in local coordinates, $\mathbf{k}_L$ = elemental stiffness in terms of local coordinates, $\mathbf{u}_L$ = displacement components expressed in local coördinates to establish the relationship:

$$\mathbf{W}_G = \mathbf{k}_G \mathbf{u}_G$$

where $\mathbf{W}_G$ = nodal force components expressed in global coordinates, $\mathbf{k}_G$ = elemental stiffness in terms of global coordinates, $\mathbf{u}_G$ = displacement components expressed in global coordinates. This is a relatively straightforward transformation provided that the relationship between $\mathbf{u}_L$ and $\mathbf{u}_G$ is known. Let us assume, at this stage, that the relationship is known and is:

$$\mathbf{u}_G = \mathbf{R}\mathbf{u}_L$$

where $\mathbf{R}$ is the associated rotational matrix.

Then, as mentioned in Chapter 7, the total strain energy stored in the system is the same, no matter what coordinates are used to define it. Hence

$$\tfrac{1}{2}(\mathbf{u}_{L1} W_{L1} + u_{L2} W_{L2} + \ldots) = \tfrac{1}{2}(u_{G1} W_{G1} + u_{G2} W_{G2} + \ldots)$$

i.e.

$$\mathbf{u}_G{}^t \mathbf{W}_G = \mathbf{u}_L{}^t \mathbf{W}_L$$

($\mathbf{u}_L$ and $\mathbf{u}_G$ are both column vectors)

$$\therefore \quad \mathbf{u}_G{}^t \mathbf{W}_G = [\mathbf{R}^{-1}\mathbf{u}_G]^t \mathbf{W}_L$$

$$\therefore \quad \mathbf{u}_G{}^t \mathbf{W}_G = \mathbf{u}_G{}^t [\mathbf{R}^{-1}]^t \mathbf{W}_L$$

$$\therefore \quad \mathbf{W}_G = [\mathbf{R}^{-1}]^t \mathbf{W}_L$$

Hence

$$\mathbf{W}_G = [\mathbf{R}^{-1}]^t\mathbf{k}_L\mathbf{u}_L = [\mathbf{R}^{-1}]^t\mathbf{k}_L\mathbf{R}^{-1}\mathbf{u}_G$$

Hence

$$\mathbf{k}_G = [\mathbf{R}^{-1}]^t\mathbf{k}_L\mathbf{R}^{-1}$$

This is a standard coordinate transformation. The rotational matrix **R** will consist of the relevant direction cosines. Let us assume that the rotation of the global axes to the local axes is such that the angle between $0X_G$ and $0X_L$ is α.

In Chapter 4 (see Coordinate transformations) equations were derived that were similar to:

$$y_1 = a_{11}x_1 + a_{21}x_2$$

$$y_2 = a_{12}x_1 + a_{22}x_2$$

In the present problem

$$x_1 = \mathbf{u}_L; \quad x_2 = \mathbf{v}_L$$

$$y_1 = \mathbf{u}_G; \quad y_2 = \mathbf{v}_G$$

The local set of coordinate axes are $0X_L$ and $0Y_L$. The global set of coordinate axes are $0X_G$ and $0Y_G$ (only the two dimensional case will be discussed here).

a_{11} = cosine of angle between $0X_L$ and $0X_G = \cos\alpha$

a_{21} = cosine of angle between $0Y_L$ and $0X_G = \cos(90° - \alpha)$

a_{12} = cosine of angle between $0X_L$ and $0Y_G = \cos(90° + \alpha)$

a_{21} = cosine of angle between $0Y_L$ and $0Y_G = \cos\alpha$

$$\therefore \quad \begin{bmatrix} u_G \\ \mathrm{v}_G \end{bmatrix} = \begin{bmatrix} \cos\alpha & \cos(90° - \alpha) \\ \cos(90° + \alpha) & \cos\alpha \end{bmatrix} \begin{bmatrix} u_L \\ \mathrm{v}_L \end{bmatrix}$$

$$= \begin{bmatrix} \cos\alpha & \sin\alpha \\ -\sin\alpha & \cos\alpha \end{bmatrix} \begin{bmatrix} u_L \\ \mathrm{v}_L \end{bmatrix}$$

If the reader is not too happy with the above proof the following pictorial representation may be of help. In Figure 8.3 the length 0A represents the actual displacement of a nodal point.

The displacements in the $0X_G$ and $0Y_G$ are u_G and v_G respectively. The displacements in the $0X_L$ and $0Y_L$ are u_L and v_L respectively. Hence, in the diagram:

$$AB = v_G; \quad 0B = u_G; \quad 0C = u_L; \quad \text{and} \quad 0E = v_L$$

$$0B \cos\alpha = 0F$$
$$= 0C + CF$$
$$= 0C + GB$$

$$\therefore \quad u_G \cos\alpha = u_L + v_G \sin\alpha$$

or

$$u_L = u_G \cos\alpha - v_G \sin\alpha \qquad \text{(A)}$$

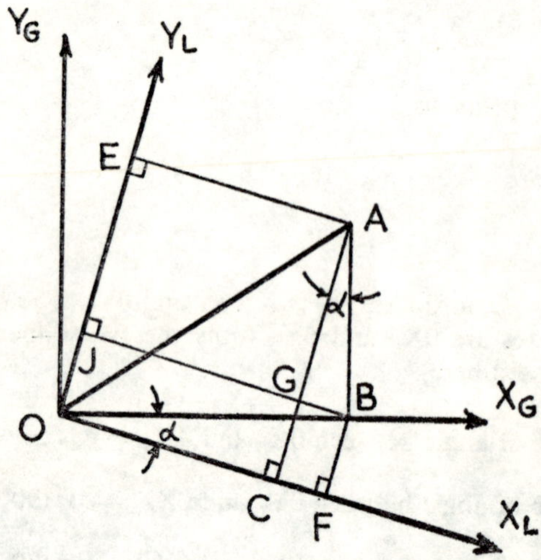

Fig. 8.3.

Also:

$$0E = 0J + JE$$
$$= BF + AG$$

$$\therefore \quad v_L = u_G \sin\alpha + v_G \cos\alpha \qquad \text{(B)}$$

From Equations (A) and (B) expressions for u_G and v_G can be found. Hence

$$\begin{bmatrix} u_G \\ v_G \end{bmatrix} = \begin{bmatrix} \cos\alpha & \sin\alpha \\ -\sin\alpha & \cos\alpha \end{bmatrix} \begin{bmatrix} u_L \\ v_L \end{bmatrix}$$

In a triangular element subjected to plane stress or strain there are six components of displacement so that the relationship $\mathbf{W}_L = \mathbf{k}_L\mathbf{u}_L$ is of the form:

$$\mathbf{W}_L = \mathbf{k}_L \begin{bmatrix} u_{L1} \\ u_{L2} \\ u_{L3} \\ v_{L1} \\ v_{L2} \\ v_{L3} \end{bmatrix}$$

where $\mathbf{k}_L$ is of course a 6×6 matrix.

Hence in order to apply the relationship $\mathbf{k}_G = \mathbf{R}^t\mathbf{k}_L\mathbf{R}$ we must form the rotational matrix $\mathbf{R}$ which will consist of a series of the submatrix

$$\begin{bmatrix} \cos\alpha & \sin\alpha \\ -\sin\alpha & \cos\alpha \end{bmatrix}$$

$$\begin{bmatrix} u_{G1} \\ v_{G1} \\ u_{G2} \\ v_{G2} \\ u_{G3} \\ v_{G3} \end{bmatrix} = \begin{bmatrix} \cos\alpha & \sin\alpha & 0 & 0 & 0 & 0 \\ -\sin\alpha & \cos\alpha & 0 & 0 & 0 & 0 \\ 0 & 0 & \cos\alpha & \sin\alpha & 0 & 0 \\ 0 & 0 & -\sin\alpha & \cos\alpha & 0 & 0 \\ 0 & 0 & 0 & 0 & \cos\alpha & \sin\alpha \\ 0 & 0 & 0 & 0 & -\sin\alpha & \cos\alpha \end{bmatrix} \begin{bmatrix} u_{L1} \\ v_{L1} \\ u_{L2} \\ v_{L2} \\ u_{L3} \\ v_{L3} \end{bmatrix}$$

$$= \mathbf{R}\mathbf{u}_L$$

It will be noticed that the displacement vector associated with local coordinate stiffness is

$$\begin{bmatrix} u_{L1} \\ u_{L2} \\ u_{L3} \\ v_{L1} \\ v_{L2} \\ v_{L3} \end{bmatrix}$$

whereas the displacement vector associated with the rotational matrix is

$$\begin{bmatrix} u_{L1} \\ v_{L1} \\ u_{L2} \\ v_{L2} \\ u_{L3} \\ v_{L3} \end{bmatrix}$$

This means that the elemental stiffness matrix, although still expressed in local coordinates, must first be amended to be in accordance with the displacement vector associated with the rotational matrix. This can quickly be achieved by the method described in Chapter 7 (See Figure 7.10). The conversion will be:

Amended stiffness matrix	1	3	5	2	4	6
Original stiffness matrix	1	2	3	4	5	6

For a rectangular element the rotational matrix, **R**, will be an 8 × 8 matrix.

THE AXIALLY SYMMETRIC CASE

Most stress problems that occur in soils can be approximated to either plane stress, plane strain or axisymmetrical stress. The classic case of an axially symmetric problem is that of a uniformly loaded circular foundation resting on soil, as illustrated in Figure 8.4A.

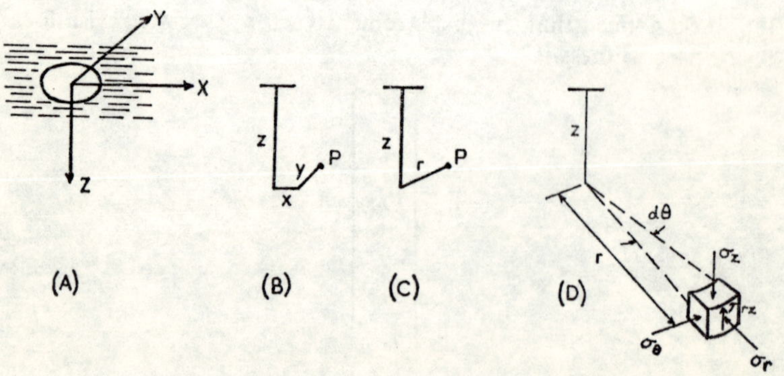

Fig. 8.4.

A point P below the foundation can be defined in Cartesian coordinates as the point (x, y, z) as in Figure 8.4B. However for a case of axial symmetry it is more convenient to define the point in polar coordinates as (r, z) where $r = \sqrt{(x^2 + y^2)}$ (Figure 8.4C).

With axial symmetry there will be no shear stresses in the vertical and radial planes. The stresses acting at a point below the foundation are:

Normal stresses:

$$\sigma_z = \text{vertical stress}$$

$$\sigma_\theta = \text{tangential stress}$$

$$\sigma_r = \text{radial stress}$$

Shear stress:

$$\tau_{rz} = \text{shear stress in the } (r, z) \text{ plane}$$

(See Figure 8.4D)

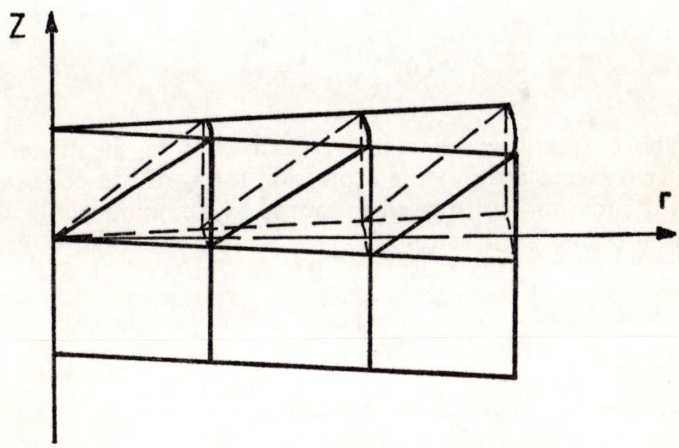

Fig. 8.5.

From the above it is evident that there can be no tangential displacements. The only displacements must be in the vertical and the radial directions. Hence the z and r axes of this case are virtually the same as the x and y coordinates of the plane stress case and the problem can be solved with the use of either rectangular, or triangular, wedge shaped elements where u becomes the radial displacement and v the vertical displacement. A typical finite element grid is illustrated in Figure 8.5.

The assumed displacement for a triangular wedged element is therefore:

$$\begin{bmatrix} u \\ \mathrm{v} \end{bmatrix} = \begin{bmatrix} 1 & r & z & 0 & 0 & 0 \\ 0 & 0 & 0 & 1 & r & z \end{bmatrix}$$

The radial displacement must induce a circumferential strain so that, unlike plane stress and plane strain, there will be a strain component in a direction perpendicular to the (r, z) plane, which is denoted as ε_θ. Hence the strains to which the element will be subjected are:

$$\varepsilon_r = \frac{\partial u}{\partial r}$$

$$\varepsilon_z = \frac{\partial \mathrm{v}}{\partial z}$$

$$\varepsilon_\theta = \frac{u}{r}$$

$$\gamma_{rz} = \frac{\partial u}{\partial z} + \frac{\partial \mathrm{v}}{\partial r}$$

The above formulae are readily obtained from the material in Chapter 6 except possibly the expression for ε_θ. If one considers an arc of a circle to be displaced outwards by an amount u then it is seen that the tangential strain,

$$\varepsilon_\theta = \frac{u}{r}$$

Hence, from the displacement functions:

$$\begin{bmatrix} \varepsilon_r \\ \varepsilon_\theta \\ \varepsilon_z \\ \gamma_{rz} \end{bmatrix} = \begin{bmatrix} 0 & 1 & 0 & 0 & 0 & 0 \\ \frac{1}{r} & 1 & \frac{z}{r} & 0 & 0 & 0 \\ 0 & 0 & 0 & 0 & 0 & 1 \\ 0 & 0 & 1 & 0 & 1 & 0 \end{bmatrix} \begin{bmatrix} \alpha_1 \\ \cdot \\ \cdot \\ \cdot \\ \alpha_6 \end{bmatrix}$$

which is of the form $\boldsymbol{\varepsilon} = \mathbf{B\alpha}$.

The elastic stress to strain matrix is of a different form to the plane stress and strain cases.

From Equation (3) of Chapter 6 we can write down.

$$\varepsilon_r = \frac{1}{E}[\sigma_r - v(\sigma_\theta + \sigma_z)]$$

$$\varepsilon_\theta = \frac{1}{E}[\sigma_\theta - v(\sigma_r + \sigma_z)]$$

$$\varepsilon_z = \frac{1}{E}[\sigma_z - v(\sigma_\theta + \sigma_r)]$$

And

$$\gamma_{rz} = \frac{1}{G}\tau_{rz} = \frac{E\tau_{rz}}{2(1 + v)}$$

Solving the above expressions for σ_r, σ_z and σ_θ eventually leads to the expression:

$$\begin{bmatrix} \sigma_r \\ \sigma_\theta \\ \sigma_z \\ \tau_{rz} \end{bmatrix} = \frac{E}{(1 + v)(1 - 2v)} \begin{bmatrix} (1 - v) & v & v & 0 \\ v & (1 - v) & v & 0 \\ v & v & (1 - v) & 0 \\ 0 & 0 & 0 & \dfrac{1 - 2v}{2} \end{bmatrix} \times \begin{bmatrix} \varepsilon_r \\ \varepsilon_\theta \\ \varepsilon_z \\ \gamma_{rz} \end{bmatrix}$$

which is of the form $\boldsymbol{\sigma} = \mathbf{D}\boldsymbol{\varepsilon}$.

EVALUATION OF STIFFNESS OF A TRIANGULAR WEDGE-SHAPED ELEMENT

Example

Suppose that the coordinates, on the centre line, of a particular element are as shown in Figure 8.6. If v is assumed to be 0·25 then the reader can, by using the techniques of Chapter 7, evaluate the generalised coordinate stiffness of the element, $\bar{\mathbf{k}}$.

The procedure is first to determine $\mathbf{B}^t\mathbf{D}\mathbf{B}$:

$$\mathbf{B}^t\mathbf{D}\mathbf{B} = 1{\cdot}6E \begin{bmatrix} \frac{3}{4r^2} & \frac{1}{r} & \frac{3z}{4r^2} & 0 & 0 & \frac{1}{4r} \\ \frac{1}{r} & 2 & \frac{z}{r} & 0 & 0 & \frac{1}{2} \\ \frac{3z}{4r^2} & \frac{z}{r} & \left(\frac{3z^2}{4r^2}+\frac{1}{4}\right) & 0 & \frac{1}{4} & \frac{z}{4r} \\ 0 & 0 & 0 & 0 & 0 & 0 \\ 0 & 0 & \frac{1}{4} & 0 & \frac{1}{4} & 0 \\ \frac{1}{4r} & \frac{1}{2} & \frac{z}{4r} & 0 & 0 & \frac{3}{4} \end{bmatrix}$$

To obtain $\bar{\mathbf{k}}$ this expression must now be integrated:

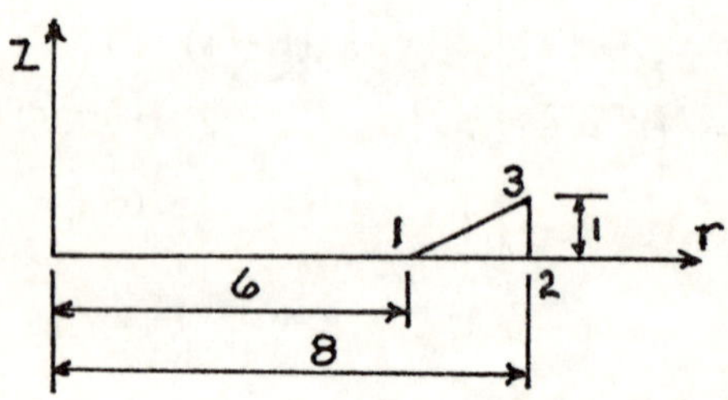

Fig. 8.6.

Hence

$$\bar{\mathbf{k}} = 2\pi \iint \mathbf{B}^t\mathbf{D}\mathbf{B}r \, dr \, dz$$

Integrating this term by term can become laborious and an approximate method, suggested by Zienkiewicz, can give reasonably accurate results. If, instead of $\mathbf{B}$, we use $\bar{\mathbf{B}}$ where $\bar{\mathbf{B}}$ is based on the coordinates of the centroidal point $\bar{r}$ and $\bar{z}$, where

$$\bar{r} = \frac{r_1 + r_2 + r_3}{3} \quad (= 22/3 \text{ in the example})$$

and

$$\bar{z} = \frac{z_1 + z_2 + z_3}{3} \quad (= 1/3 \text{ in the example})$$

then $\bar{\mathbf{k}}$ is simply $2\pi\bar{\mathbf{B}}^t\mathbf{D}\bar{\mathbf{B}}r\Delta$, where Δ = the area of the triangle

(= 1·0 in the example). Hence, with this approximation:

$$\bar{\mathbf{k}} = 1{\cdot}6E \begin{bmatrix} 0{\cdot}0140 & 0{\cdot}1364 & 0{\cdot}0047 & 0{\cdot}0000 & 0{\cdot}0000 & 0{\cdot}0341 \\ 0{\cdot}1364 & 2{\cdot}0000 & 0{\cdot}0455 & 0{\cdot}0000 & 0{\cdot}0000 & 0{\cdot}5000 \\ 0{\cdot}0047 & 0{\cdot}0455 & 0{\cdot}2516 & 0{\cdot}0000 & 0{\cdot}2500 & 0{\cdot}0114 \\ 0{\cdot}0000 & 0{\cdot}0000 & 0{\cdot}0000 & 0{\cdot}0000 & 0{\cdot}0000 & 0{\cdot}0000 \\ 0{\cdot}0000 & 0{\cdot}0000 & 0{\cdot}2500 & 0{\cdot}0000 & 0{\cdot}2500 & 0{\cdot}0000 \\ 0{\cdot}0341 & 0{\cdot}5000 & 0{\cdot}0114 & 0{\cdot}0000 & 0{\cdot}0000 & 0{\cdot}7500 \end{bmatrix}$$

The coordinate elemental stiffness is found as before from

$$\mathbf{k} = [\mathbf{A}^{-1}]^t \bar{\mathbf{k}} \mathbf{A}^{-1}$$

where

$$\mathbf{A} = \begin{bmatrix} 1 & 6 & 0 & 0 & 0 & 0 \\ 1 & 8 & 0 & 0 & 0 & 0 \\ 1 & 8 & 1 & 0 & 0 & 0 \\ 0 & 0 & 0 & 1 & 6 & 0 \\ 0 & 0 & 0 & 1 & 8 & 0 \\ 0 & 0 & 0 & 1 & 8 & 1 \end{bmatrix}$$

RECTANGULAR ELEMENTS

Displacement functions similar to those used above can be evolved for elements rectangular in the (r, z) plane (Clough, 1965). In this case the displacement matrix becomes:

$$\begin{bmatrix} u \\ v \end{bmatrix} = \begin{bmatrix} 1 & r & z & rz & 0 & 0 & 0 & 0 \\ 0 & 0 & 0 & 0 & 1 & r & z & rz \end{bmatrix} \begin{bmatrix} \alpha_1 \\ \cdot \\ \cdot \\ \cdot \\ \alpha_8 \end{bmatrix}$$

and **B** becomes:

$$\mathbf{B} = \begin{bmatrix} 0 & 1 & 0 & z & 0 & 0 & 0 & 0 \\ \frac{1}{r} & 1 & \frac{z}{r} & z & 0 & 0 & 0 & 0 \\ 0 & 0 & 0 & 0 & 0 & 0 & 1 & r \\ 0 & 0 & 1 & r & 0 & 1 & 0 & z \end{bmatrix}$$

FURTHER READING

The material in this chapter is intended to give some examples of the further application of the finite element method. An interested reader who wishes to progress further will find it necessary to obtain some of the further reading matter listed in the Appendix under Bibliography.

Appendix I

Solution of Simultaneous Equations by Cramer's Rule

IF two variables, x and y, are related by a set of simultaneous equations of the form:

$$
\begin{aligned}
y_1 &= a_{11}x_1 + a_{12}x_2 + a_{13}x_3 + \dots + a_{1n}x_n \\
y_2 &= a_{21}x_1 + a_{22}x_2 + a_{23}x_3 + \dots + a_{2n}x_n \\
\cdot \quad & \quad \cdot \qquad \cdot \qquad \cdot \qquad \dots \quad \cdot \\
\cdot \quad & \quad \cdot \qquad \cdot \qquad \cdot \qquad \dots \quad \cdot \\
\cdot \quad & \quad \cdot \qquad \cdot \qquad \cdot \qquad \dots \quad \cdot \\
y_n &= a_{n1}x_1 + a_{n2}x_2 + a_{n3}x_3 + \dots + a_{nn}x_n
\end{aligned}
$$

then the relationships can be expressed in matrix form:

$$\mathbf{y} = \mathbf{A}\mathbf{x}$$

where

$$
\mathbf{y} = \begin{bmatrix} y_1 \\ y_2 \\ y_3 \\ \cdot \\ \cdot \\ \cdot \\ y_n \end{bmatrix}
\quad
\mathbf{A} = \begin{bmatrix} a_{11} & a_{12} & a_{13} & \dots & a_{1n} \\ a_{21} & a_{22} & a_{23} & \dots & a_{2n} \\ \cdot & \cdot & \cdot & \dots & \cdot \\ \cdot & \cdot & \cdot & \dots & \cdot \\ \cdot & \cdot & \cdot & \dots & \cdot \\ a_{n1} & a_{n2} & a_{n3} & \dots & a_{nn} \end{bmatrix}
\quad
\mathbf{x} = \begin{bmatrix} x_1 \\ x_2 \\ x_3 \\ \cdot \\ \cdot \\ \cdot \\ x_n \end{bmatrix}
$$

The reverse transformation is $\mathbf{x} = \mathbf{A}^{-1}\mathbf{y}$ where $\mathbf{A}^{-1}$ is the inverse of $\mathbf{A}$. This expression may be written as:

$$\mathbf{x} = \frac{\mathbf{A}}{|\mathbf{A}|}\,\mathbf{y}$$

where $\mathbf{A}^{a}$ is the adjoint matrix, the transpose of the cofactor matrix $\mathbf{A}^{c}$.

Now $|\mathbf{A}|$ is the determinant of matrix $\mathbf{A}$. i.e.

$$\begin{bmatrix} x_1 \\ x_2 \\ \cdot \\ \cdot \\ \cdot \\ x_n \end{bmatrix} = \frac{1}{|\mathbf{A}|} \begin{bmatrix} a_{11}^c & a_{21}^c & \dots & a_{n1}^c \\ a_{12}^c & a_{22}^c & \dots & a_{n2}^c \\ \cdot & \cdot & \dots & \cdot \\ \cdot & \cdot & \dots & \cdot \\ \cdot & \cdot & \dots & \cdot \\ a_{1n}^c & a_{2n}^c & \dots & a_{nn}^c \end{bmatrix} \begin{bmatrix} y_1 \\ y_2 \\ \cdot \\ \cdot \\ \cdot \\ y_n \end{bmatrix}$$

Consider the first term x_{11}

$$x_{11} = \frac{y_1 a_{11}^c + y_2 a_{21}^c + \dots + y_n a_{n1}^c}{|\mathbf{A}|}$$

The numerator is seen to be the expansion of the cofactors of the following determinant:

$$\begin{vmatrix} y_1 & a_{12} & a_{13} & \dots & a_{1n} \\ y_2 & a_{22} & a_{23} & \dots & a_{2n} \\ y_3 & a_{32} & a_{33} & \dots & a_{3n} \\ \cdot & \cdot & \cdot & \dots & \cdot \\ \cdot & \cdot & \cdot & \dots & \cdot \\ \cdot & \cdot & \cdot & \dots & \cdot \\ y_n & a_{n2} & a_{n3} & \dots & a_{nn} \end{vmatrix}$$

Similarly

$$x_2 = \frac{y_1 a_{12}^c + y_2 a_{22}^c + \dots + y_n a_{n2}^c}{|\mathbf{A}|}$$

The numerator being:

$$\begin{vmatrix} a_{11} & y_1 & a_{13} & \dots & a_{1n} \\ a_{21} & y_2 & a_{23} & \dots & a_{2n} \\ a_{31} & y_3 & a_{33} & \dots & a_{3n} \\ \cdot & \cdot & \cdot & \dots & \cdot \\ \cdot & \cdot & \cdot & \dots & \cdot \\ \cdot & \cdot & \cdot & \dots & \cdot \\ a_{n1} & y_n & a_{n3} & \dots & a_{nn} \end{vmatrix}$$

In each expression the denominator is always the determinant $|\mathbf{A}|$ whereas the numerator is equal to $|\mathbf{A}|$ with one column replaced by the components of the vector y. The number of the column that is replaced by y is the number of the unknown x term, i.e. the first column is replaced for x_1, the second column is replaced for x_2, the n-th column is replaced for x_n.

This is Cramer's rule. The application of this rule is illustrated in Chapter 4.

Appendix II
Roots of a Cubic Equation

IN determining the principal stresses of a triaxial system one has to solve the characteristic equation of the system which can be expressed in terms of the stress invariants:

$$\sigma^3 - J_1\sigma^2 + J_2\sigma - J_3 = 0$$

where

$$J_1 = \sigma_1 + \sigma_2 + \sigma_3$$

$$J_2 = \sigma_1\sigma_2 + \sigma_2\sigma_3 + \sigma_3\sigma_1$$

$$J_3 = \sigma_1\sigma_2\sigma_3$$

where σ_1, σ_2 and σ_3 are the principal stresses of the system. If σ' is put equal to

$$\sigma - \frac{J_1}{3}$$

i.e.

$$\sigma = \sigma' + \frac{J_1}{3}$$

then:

$$\left(\sigma'^3 + \frac{3J_1\sigma'^2}{3} + \frac{3}{9}\sigma' J_1{}^2 + \frac{J_1}{27}\right) - J_1\left(\sigma'^2 + \frac{2\sigma' J_1}{3} + \frac{J_1{}^2}{9}\right)$$

$$+ J_2\left(\sigma' + \frac{J_1}{3}\right) - J_3 = 0$$

i.e.

$$\sigma'^3 - \left(\frac{J_1{}^2}{3} - J_2\right)\sigma' - \left(\frac{2J_1{}^3}{27} - \frac{J_1J_2}{3} + J_3\right) = 0$$

or

$$\sigma'^3 - J_2'\sigma' - J_3' = 0$$

where

$$J_2' = \left(\frac{J_1{}^2}{3} - J_2\right)$$

$$J_3' = \left(\frac{2J_1{}^3}{27} - \frac{J_1J_2}{3} + J_3\right)$$

If we express σ' as $r \cos \theta$ the expression becomes:

$$r^3 \cos^3 \theta - J_2' r \cos \theta - J_3' = 0$$

or

$$\cos^3 \theta - \frac{J_2'}{r^2} \cos \theta - \frac{J_3'}{r^3} = 0$$

which is identical to the trigonometrical expression

$$\cos^3 \theta - \frac{3}{4} \cos \theta - \frac{1}{4} \cos 3\theta = 0$$

when

$$\frac{J_2'}{r^2} = \frac{3}{4}; \quad \frac{J_3'}{r^3} = \frac{\cos 3\theta}{4}$$

i.e.

$$r = \sqrt{\left(\frac{4J_2'}{3}\right)}$$

$$\cos 3\theta = \frac{4J_3'}{r^3}$$

Example

If the characteristic equation is

$$\sigma^3 - 34\sigma^2 + 291\sigma - 584 = 0$$

Then

$$J_1 = 34; \quad J_2 = 291; \quad J_3 = 584$$

$$J_2' = \left(\frac{J_1^2}{3} - J_2\right) = \left(\frac{34^2}{3} - 291\right) = 94{\cdot}3$$

$$J_3' = \left(\frac{2J_1^3}{27} - \frac{J_1 J_2}{3} + J_3\right) = \left(\frac{2 \times 34^3}{27} - \frac{34 \times 291}{3} + 584\right)$$

$$= 2912 - 3298 + 584 = 198$$

Now

$$\sigma'^3 - J_2'\sigma'^2 - J_3' = 0$$

$$\therefore \quad \sigma'^3 - 94{\cdot}3\sigma'^2 - 198 = 0$$

$$r = \sqrt{\left(\frac{4 \times 94{\cdot}3}{3}\right)} = 11{\cdot}21; \quad \cos 3\theta = \frac{4 \times 198}{11{\cdot}21^3} = 0{\cdot}5623$$

$$3\theta = 55°47' \qquad 304°13' \qquad 415°47'$$

Therefore

$$\theta = 18°35' \qquad 101°24' \qquad 138°36'$$

$$\cos\theta = 0{\cdot}9479 \qquad -0{\cdot}1977 \qquad -0{\cdot}7501$$

$$\sigma' = r\cos\theta = 10{\cdot}64 \qquad -2{\cdot}216 \qquad -8{\cdot}41$$

$$\sigma = \sigma' + \frac{J_1}{3} = 21{\cdot}97 \qquad 9{\cdot}11 \qquad 2{\cdot}92$$

Appendix III

Derivation of

$$\frac{\partial^4 \varphi}{\partial x^4} + 2\frac{\partial^4 \varphi}{\partial x^2 \partial z^2} + \frac{\partial^4 \varphi}{\partial z^4} = 0$$

AS was stated in Chapter 6 the two dimensional state of stress in an elastic body when weight is the only significant body force can be represented by the above differential equation. Proof of this statement is not necessary in order to understand the contents of the book but, for those interested, the derivation is set out in the following pages.

EQUILIBRIUM EQUATIONS

All stresses vary from point to point within a stressed system and, although elsewhere in the text they have been assumed equal, the stresses on opposite faces of an elemental cube are slightly different. This fact will now be examined by studying the equilibrium of such an elemental cube. As the various stress differences are small the body forces acting on the element must also be included as they are of the same order of magnitude. Let us consider the two dimensional stress system shown in Figure A.1. In this figure the body force components are represented by X and Z. As the various faces of the element are small we can assume that the corresponding forces arising from the applied stresses are equal to the product of the particular stress and the area of the face over which it acts. These forces must be in equilibrium and considering them as positive when acting in the negative direction of their respective axes and summing forces in the x direction:

$$\left(\sigma_x + \frac{\partial \sigma_x}{\partial x}\,\mathrm{d}x - \sigma_x\right)\mathrm{d}x\,\mathrm{d}z + \left(\tau_{zx} + \frac{\partial \tau_{zx}}{\partial z}\,\mathrm{d}z - \tau_{zx}\right)\mathrm{d}x\,\mathrm{d}y$$
$$+\, x\,\mathrm{d}x\,\mathrm{d}y\,\mathrm{d}z = 0$$

i.e.

$$X\,\mathrm{d}x\,\mathrm{d}y\,\mathrm{d}z + \frac{\partial \sigma_x}{\partial x}\,\mathrm{d}x\,\mathrm{d}y\,\mathrm{d}z + \frac{\partial \tau_{zx}}{\partial z}\,\mathrm{d}x\,\mathrm{d}y\,\mathrm{d}z = 0$$

i.e.

$$\frac{\partial \sigma_x}{\partial x} + \frac{\partial \tau_{zx}}{\partial z} + X = 0 \qquad (1)$$

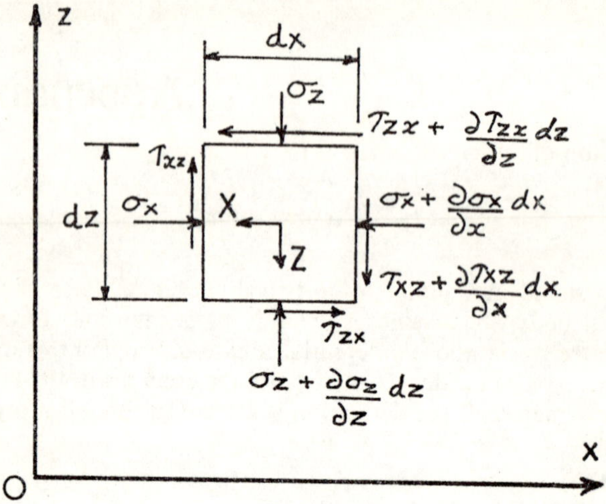

Fig. A.1.

Similarly for forces in the z direction:

$$\frac{\partial \sigma_z}{\partial z} + \frac{\partial \tau_{zx}}{\partial x} + Z = 0 \tag{2}$$

COMPATIBILITY EQUATIONS

Generally in stress analysis we are concerned with the problem of determining the state of stress in a body, given the applied loading and the boundary conditions. For example if the body has an unloaded surface then, along this surface there can be neither normal nor shear stresses acting.

The equilibrium equations (1 and 2) were derived on the assumption that the elemental cube was an absolutely rigid body. Therefore with the application of statics and a knowledge of the boundary conditions it is still not possible to evaluate σ_x, σ_z and τ_{zx} unless we also consider the elastic deformation of the body. The problem is statically indeterminate and its solution must involve the formulation of a set of equations that express the condition that the state of stress at a point in the mass is compatible with the elastic properties of the material.

Expressions for the six strain components have been derived in

Chapter 6. For a two dimensional problem only three strain components are required.

$$\varepsilon_x = \frac{\partial x}{\partial u}; \quad \varepsilon_z = \frac{\partial w}{\partial z}; \quad \gamma_{xz} = \frac{\partial u}{\partial z} + \frac{\partial w}{\partial x}$$

These three components are all in terms of the functions u and w and some relationship must therefore exist between them. If we differentiate:

(i) ε_x twice with respect to z

(ii) ε_z twice with respect to x

(iii) γ_{xz} with respect to x and with respect to z:

$$\frac{\partial^2 \varepsilon_x}{\partial z^2} = \frac{\partial^3 u}{\partial x \partial z^2}; \quad \frac{\partial^2 \varepsilon_z}{\partial x^2} = \frac{\partial^3 w}{\partial z \partial x^2}; \quad \frac{\partial^2 \gamma_{xz}}{\partial x \partial z} = \frac{\partial^2}{\partial x \partial z}\left(\frac{\partial w}{\partial x} + \frac{\partial u}{\partial z}\right)$$

Then u and w can be eliminated from these terms and the following expression developed:

$$\frac{\partial^2 \varepsilon_x}{\partial z^2} + \frac{\partial^2 \varepsilon_z}{\partial x^2} = \frac{\partial^2 \gamma_{xz}}{\partial x \partial z} \tag{3}$$

This is the compatibility equation in terms of strain. For a three dimensional problem there would be six of these equations. The equation can be expressed in terms of stress by the use of Equations (3) and (4) of Chapter 6:

$$\varepsilon_x = \frac{\sigma_x}{E} - \frac{v\sigma_z}{E}$$

(In plane stress $\sigma_y = 0$)

$$\varepsilon_z = \frac{\sigma_z}{E} - \frac{v\sigma_x}{E}$$

$$\gamma_{zx} = \frac{\tau_{zx}}{G} = 2\tau_{zx}\frac{(1 + v)}{E}$$

Substituting in Equation (3):

$$\frac{\partial^2}{\partial z^2}(\sigma_x - v\sigma_z) + \frac{\partial^2}{\partial x^2}(\sigma_z - v\sigma_x) = 2(1 + v)\frac{\partial^2 \tau_{zx}}{\partial x \partial z} \tag{4}$$

By making use of the equilibrium equations the above expression can be written in a different form. In soil problems the only significant body force is quite often the weight. When this is the case the equilibrium equations (1 and 2) become:

$$\frac{\partial \sigma_x}{\partial x} + \frac{\partial \tau_{zx}}{\partial z} = 0 \tag{5}$$

$$\frac{\partial \sigma_z}{\partial z} + \frac{\partial \tau_{zx}}{\partial x} + Z = 0 \tag{6}$$

Differentiating the first of these expressions with respect to x and the second with respect to z and then adding gives:

$$2\frac{\partial^2 \tau_{zx}}{\partial x \partial z} = -\frac{\partial^2 \sigma_x}{\partial x^2} - \frac{\partial^2 \sigma_z}{\partial z^2}$$

and, substituting in Equation (4) the compatibility equation expressed in stress components, becomes:

$$\left(\frac{\partial^2}{\partial x^2} + \frac{\partial^2}{\partial z^2}\right)(\sigma_x + \sigma_z) = 0 \tag{7}$$

In the case of constant body forces the equations that determine the stress distribution do not contain the elastic constants E and v. The stress distribution is therefore the same for all isotropic materials, a fact that is made use of in polarised light systems where, by the use of a transparent material such as glass or perspex, it is possible to determine a stress system experimentally.

STRESS FUNCTIONS

It has been pointed out that the solution of a stress problem is dependent upon the integration of the differential equation of equilibrium together with the boundary conditions and the compatibility equation.

For a two dimensional stress system, when weight is the only significant body force the equations to be solved are (5, 6 and 7). The solution of these equations can be obtained by introducing a new term known as a stress function or the Airy stress function. A stress function is simply a mathematical function of the two variables x and z in which the stress components are expressed in terms of this new function.

Equations (5 and 6) are satisfied if we take a function of x and z, φ, such that:

$$\sigma_x = \frac{\partial^2 \varphi}{\partial z^2}; \quad \sigma_z = \frac{\partial^2 \varphi}{\partial x^2}; \quad \tau_{zx} = -\frac{\partial^2 \varphi}{\partial x \partial z}$$

Such a function will lead to a variety of solutions of the equilibrium equations. The correct solution will be the one that also satisfies the compatibility Equation (7). Substituting our expressions for σ_x, σ_z and τ_{zx} into the equation gives:

$$\frac{\partial^4 \varphi}{\partial x^4} + 2\frac{\partial^4 \varphi}{\partial x^2 \partial z^2} + \frac{\partial^4 \varphi}{\partial z^4} = 0$$

and the solution of a two dimensional stress problem is reduced to finding a solution for this equation which satisfies the boundary conditions.

Appendix IV

Computers

THE methods described in this book can only be used in conjunction with a computer system. In order to computerise a particular problem it is first necessary to prepare a list of operational steps that the computer will have to carry out in order for it to give the required answers. Such a set of steps is known as a flow diagram.

Having prepared the necessary flow diagram it must then be converted into a list of instructions written in a language that will be accepted by the computer. The list is referred to as the program. A computer will accept a program written in a variety of languages but engineers and students normally use special high-level languages which have been designed with their requirements in mind. Two such languages are Algol 60 and Fortran IV and are widely used in modern scientific and engineering applications.

A knowledge of one or other is all that is required as, with certain local conditions to suit a particular computer, a program written in either language will be accepted by most modern digital machines. Many excellent books dealing with computers are available but for a reader interested in this field the following publications, obtainable from Heriot–Watt University, are both cheap and easy to follow. *Algol programming notes*, by A. Balfour and C. J. Crook. *An introduction to Fortran IV programming*, by A. Balfour and A. J. McTernan. These publications are in the form of duplicated notes and may be obtained from

The Secretary,
Department of Computer Science,
Heriot–Watt University,
37/39 Grassmarket,
Edinburgh, EH1 2HW.

COMPUTER KNOWLEDGE REQUIRED BY THE PRACTISING ENGINEER

For the present day student a knowledge of computer programming is almost obligatory, but for the practising engineer with little time to spare, there are facilities available to him by which he can have his problems analysed by computer without becoming directly involved himself. In spite of this, some knowledge is always an advantage to the engineer in that he can then talk the same language as the people who will deal with his problem.

Many of the problems that occur in civil engineering have been fully examined and there are standard programs available for their solution. Such programs are a normal part of the services offered by computer firms to their clients. However, for topics covered in this book, there are few programs available although most firms would willingly undertake to program such a project.

COMPUTER BUREAUX

The simplest way for an engineer to obtain advice on computer techniques is for him to pick up the classified telephone directory in which he will find the names and addresses of the local computer bureaux. These bureaux are commercial organisations who sell time and expertise geared to computer configuration.

In most directories the number of firms listed is large and there is the risk that the enquirer may arbitrarily contact a firm that is not ideally suited to deal with his particular problem.

COMPUTER FIRMS

A more direct approach is to contact a computer manufacturing firm. These firms, besides selling their machines, have contact with bureaux that specialise in their equipment and can therefore not only give advice on programming matters, etc. but can also put the enquirer into contact with a suitable local bureau.

DIRECT ACCESS TO A COMPUTER

The cost of a computer installation is such that only a large organisation can afford the luxury of their own machine. Nevertheless it is possible for any engineer to have direct access to a machine if he wants to. Most computer bureaux offer a service whereby a customer can hire their machine and feed in his own programs.

A multi-access time sharing system is available on most of the big computers available in this country. With this system, under agreement with a computer firm, a remote terminal unit, i.e. teleprinter etc., is installed in the customer's own office so that, with the aid of the GPO lines, he is able to dial his particular computer and 'speak' to it whenever he wishes. Firms that provide this service also organise courses so that their customers become familiar with the facilities available.

Most universities and some technical colleges offer a computer service to the general public by which engineers are allowed direct access to their machine.

Appendix V

Explicit Form of Elemental Stiffness Matrix

THE numerical methods described in Chapters 7 and 8 that are necessary to evolve the elemental stiffness matrix are best carried out in a computer.

However there are occasions, particularly in the early development of a computer program, when numerical checks on the various matrices involved can be useful.

Formulae can be evolved by which the stiffness matrices for standard elements, triangular, rectangular, etc., can be determined by simply inserting the various coordinates of the elemental nodes.

Set out below is the derivation of the explicit form of the stiffness matrix of a triangular element, in either plane stress or plane strain.

Consider the triangular element shown in Figure A.2A.

In its general form the matrix **A** will be:

$$\mathbf{A} = \begin{bmatrix} 1 & x_1 & y_1 & 0 & 0 & 0 \\ 1 & x_2 & y_2 & 0 & 0 & 0 \\ 1 & x_3 & y_3 & 0 & 0 & 0 \\ 0 & 0 & 0 & 1 & x_1 & y_1 \\ 0 & 0 & 0 & 1 & x_2 & y_2 \\ 0 & 0 & 0 & 1 & x_3 & y_3 \end{bmatrix}$$

We require to find $\mathbf{A}^{-1}$ but, if we partition the matrix, we need only find the inverse of the matrix

$$\mathbf{A}_{11} = \begin{bmatrix} 1 & x_1 & y_1 \\ 1 & x_2 & y_2 \\ 1 & x_3 & y_3 \end{bmatrix}$$

The reader can quickly verify that:

$$\mathbf{A}_{11}{}^{a} = \begin{bmatrix} x_2y_3 - x_3y_2 & x_3y_1 - x_1y_3 & x_1y_2 - x_2y_1 \\ y_2 - y_3 & y_3 - y_1 & y_1 - y_2 \\ x_3 - x_2 & x_1 - x_3 & x_2 - x_1 \end{bmatrix}$$

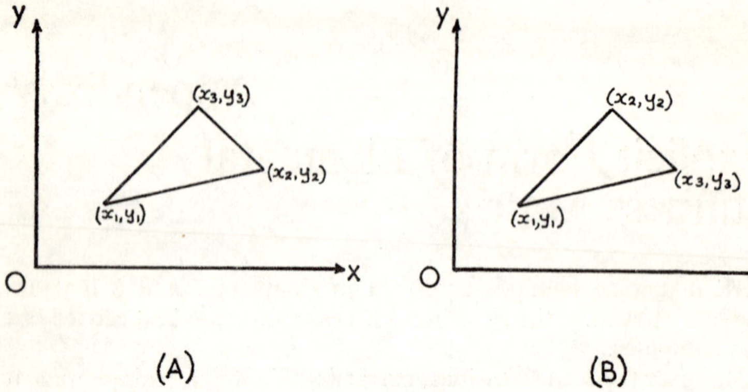

Fig. A.2.

and that:

$$|\mathbf{A}_{11}| = x_1(y_2 - y_3) + x_2(y_3 - y_1) + x_3(y_1 - y_2) = 2\Delta$$

where Δ = the area of the element. Hence, as

$$\mathbf{A}_{11}{}^{-1} = \frac{\mathbf{A}_{11}{}^{a}}{|\mathbf{A}_{11}|}$$

$$\mathbf{A}^{-1} = \frac{1}{2\Delta}\begin{bmatrix} x_2y_3 - x_3y_2 & x_3y_1 - x_1y_3 & x_1y_2 - x_2y_1 & 0 & 0 & 0 \\ y_2 - y_3 & y_3 - y_1 & y_1 - y_2 & 0 & 0 & 0 \\ x_3 - x_2 & x_1 - x_3 & x_2 - x_1 & 0 & 0 & 0 \\ 0 & 0 & 0 & x_2y_3 - x_3y_2 & x_3y_1 - x_1y_3 & x_1y_2 - x_2y_1 \\ 0 & 0 & 0 & y_2 - y_3 & y_3 - y_1 & y_1 - y_2 \\ 0 & 0 & 0 & x_3 - x_2 & x_1 - x_3 & x_2 - x_1 \end{bmatrix}$$

Note. In Figure A.2A the nodes of the element were lettered in an anticlockwise direction. If the element had been lettered as shown in

Figure A.2B (i.e. clockwise) the determinant of **A** then represents the negative value of twice the area of the element and a negative sign would therefore have to be placed in front of the expression for matrix $\mathbf{A}^{-1}$.

If we represent the stress–strain matrix **D** in the form

$$\mathbf{D} = \begin{bmatrix} d_{11} & d_{12} & 0 \\ d_{21} & d_{22} & 0 \\ 0 & 0 & d_{33} \end{bmatrix}$$

where

$$d_{11} = d_{22} = \frac{E}{1 - v^2} \text{ (for plane stress)} = \frac{E(1 - v)}{(1 + v)(1 - 2v)} \text{ (for plane strain)}$$

$$d_{12} = d_{21} = \frac{vE}{1 - v^2} \text{ (for plane stress)} = \frac{vE}{(1 + v)(1 - 2v)} \text{ (for plane strain)}$$

$$d_{33} = \frac{E}{2(1 + v)} \text{ (for plane stress)} = \frac{E}{2(1 + v)} \text{ (for plane strain)}$$

then we can obtain a general expression for $\bar{\mathbf{k}}$

$$\bar{\mathbf{k}} = \Delta T \mathbf{B}^t \mathbf{D} \mathbf{B}$$

$$= \Delta T \begin{bmatrix} 0 & 0 & 0 \\ 1 & 0 & 0 \\ 0 & 0 & 1 \\ 0 & 0 & 0 \\ 0 & 0 & 1 \\ 0 & 1 & 0 \end{bmatrix} \begin{bmatrix} d_{11} & d_{12} & 0 \\ d_{21} & d_{22} & 0 \\ 0 & 0 & d_{33} \end{bmatrix} \begin{bmatrix} 0 & 1 & 0 & 0 & 0 & 0 \\ 0 & 0 & 0 & 0 & 0 & 1 \\ 0 & 0 & 1 & 0 & 1 & 0 \end{bmatrix}$$

$$= \Delta T \begin{bmatrix} 0 & 0 & 0 & 0 & 0 & 0 \\ 0 & d_{11} & 0 & 0 & 0 & d_{12} \\ 0 & 0 & d_{33} & 0 & d_{33} & 0 \\ 0 & 0 & 0 & 0 & 0 & 0 \\ 0 & 0 & d_{33} & 0 & d_{33} & 0 \\ 0 & d_{21} & 0 & 0 & 0 & d_{22} \end{bmatrix}$$

Now

$$\mathbf{k} = [\mathbf{A}^{-1}]^t \bar{\mathbf{k}} \mathbf{A}^{-1}$$

When these final matrix multiplications are carried out we obtain the explicit form of the stiffness matrix of a triangular element as shown in Figure A.3.

Notes

1. The matrix is symmetrical.
2. In the final formula the fact that the elemental nodes have been lettered clockwise or anti-clockwise is of no consequence as two negative, or two positive, values multiplied together give a positive product.
3. The values corresponding to plane stress or plane strain must be inserted into the formula for d_{11}, d_{12}, etc.

$$
\mathbf{k} = \frac{\mathbf{T}}{4\Delta}
\begin{bmatrix}
\begin{matrix} d_{11}(y_2 - y_3)^2 \\ + \\ d_{33}(x_3 - x_2)^2 \end{matrix} &
\begin{matrix} d_{11}(y_2 - y_3)(y_3 - y_1) \\ + \\ d_{33}(x_3 - x_2)(x_1 - x_3) \end{matrix} &
\begin{matrix} d_{11}(y_2 - y_3)(y_1 - y_2) \\ + \\ d_{33}(x_3 - x_2)(x_2 - x_1) \end{matrix} &
\begin{matrix} d_{33}(x_3 - x_2)(y_2 - y_3) \\ + \\ d_{21}(x_3 - x_2)(y_2 - y_3) \end{matrix} &
\begin{matrix} d_{33}(x_3 - x_2)(y_3 - y_1) \\ + \\ d_{21}(x_1 - x_3)(y_2 - y_3) \end{matrix} &
\begin{matrix} d_{33}(x_3 - x_2)(y_1 - y_2) \\ + \\ d_{21}(x_2 - x_1)(y_2 - y_3) \end{matrix} \\
\begin{matrix} d_{11}(y_2 - y_3)(y_3 - y_1) \\ + \\ d_{33}(x_3 - x_2)(x_1 - x_3) \end{matrix} &
\begin{matrix} d_{11}(y_3 - y_1)^2 \\ + \\ d_{33}(x_1 - x_3)^2 \end{matrix} &
\begin{matrix} d_{11}(y_3 - y_1)(y_1 - y_2) \\ + \\ d_{33}(x_1 - x_3)(x_2 - x_1) \end{matrix} &
\begin{matrix} d_{33}(x_1 - x_3)(y_2 - y_3) \\ + \\ d_{21}(x_3 - x_2)(y_3 - y_1) \end{matrix} &
\begin{matrix} d_{33}(x_1 - x_3)(y_3 - y_1) \\ + \\ d_{21}(x_1 - x_3)(y_3 - y_1) \end{matrix} &
\begin{matrix} d_{33}(x_1 - x_3)(y_1 - y_2) \\ + \\ d_{21}(x_2 - x_1)(y_3 - y_1) \end{matrix} \\
\begin{matrix} d_{11}(y_2 - y_3)(y_1 - y_2) \\ + \\ d_{33}(x_3 - x_2)(x_2 - x_1) \end{matrix} &
\begin{matrix} d_{11}(y_3 - y_1)(y_1 - y_2) \\ + \\ d_{33}(x_1 - x_3)(x_2 - x_1) \end{matrix} &
\begin{matrix} d_{11}(y_1 - y_2)^2 \\ + \\ d_{33}(x_2 - x_1)^2 \end{matrix} &
\begin{matrix} d_{33}(x_2 - x_1)(y_2 - y_3) \\ + \\ d_{21}(x_3 - x_2)(y_1 - y_2) \end{matrix} &
\begin{matrix} d_{33}(x_2 - x_1)(y_3 - y_1) \\ + \\ d_{21}(x_1 - x_3)(y_1 - y_2) \end{matrix} &
\begin{matrix} d_{33}(x_2 - x_1)(y_1 - y_2) \\ + \\ d_{21}(x_2 - x_1)(y_1 - y_2) \end{matrix} \\
\begin{matrix} d_{33}(x_3 - x_2)(y_2 - y_3) \\ + \\ d_{21}(x_3 - x_2)(y_2 - y_3) \end{matrix} &
\begin{matrix} d_{33}(x_1 - x_3)(y_2 - y_3) \\ + \\ d_{21}(x_3 - x_2)(y_3 - y_1) \end{matrix} &
\begin{matrix} d_{33}(x_2 - x_1)(y_2 - y_3) \\ + \\ d_{21}(x_3 - x_2)(y_1 - y_2) \end{matrix} &
\begin{matrix} d_{33}(y_2 - y_3)^2 \\ + \\ d_{22}(x_3 - x_2)^2 \end{matrix} &
\begin{matrix} d_{33}(y_2 - y_3)(y_3 - y_1) \\ + \\ d_{22}(x_3 - x_2)(x_1 - x_3) \end{matrix} &
\begin{matrix} d_{33}(y_2 - y_3)(y_1 - y_2) \\ + \\ d_{22}(x_3 - x_2)(x_2 - x_1) \end{matrix} \\
\begin{matrix} d_{33}(x_3 - x_2)(y_3 - y_1) \\ + \\ d_{21}(x_1 - x_3)(y_2 - y_3) \end{matrix} &
\begin{matrix} d_{33}(x_1 - x_3)(y_3 - y_1) \\ + \\ d_{21}(x_1 - x_3)(y_3 - y_1) \end{matrix} &
\begin{matrix} d_{33}(x_2 - x_1)(y_3 - y_1) \\ + \\ d_{21}(x_1 - x_3)(y_1 - y_2) \end{matrix} &
\begin{matrix} d_{33}(y_2 - y_3)(y_3 - y_1) \\ + \\ d_{22}(x_3 - x_2)(x_1 - x_3) \end{matrix} &
\begin{matrix} d_{33}(y_3 - y_1)^2 \\ + \\ d_{22}(x_1 - x_3)^2 \end{matrix} &
\begin{matrix} d_{33}(y_3 - y_1)(y_1 - y_2) \\ + \\ d_{22}(x_1 - x_3)(x_2 - x_1) \end{matrix} \\
\begin{matrix} d_{33}(x_3 - x_2)(y_1 - y_2) \\ + \\ d_{21}(x_2 - x_1)(y_2 - y_3) \end{matrix} &
\begin{matrix} d_{33}(x_1 - x_3)(y_1 - y_2) \\ + \\ d_{21}(x_2 - x_1)(y_3 - y_1) \end{matrix} &
\begin{matrix} d_{33}(x_2 - x_1)(y_1 - y_2) \\ + \\ d_{21}(x_2 - x_1)(y_1 - y_2) \end{matrix} &
\begin{matrix} d_{33}(y_2 - y_3)(y_1 - y_2) \\ + \\ d_{22}(x_3 - x_2)(x_2 - x_1) \end{matrix} &
\begin{matrix} d_{33}(y_3 - y_1)(y_1 - y_2) \\ + \\ d_{22}(x_1 - x_3)(x_2 - x_1) \end{matrix} &
\begin{matrix} d_{33}(y_1 - y_2)^2 \\ + \\ d_{22}(x_2 - x_1)^2 \end{matrix}
\end{bmatrix}
$$

Fig. A.3. *Explicit form of the stiffness matrix of a triangular element.*

Appendix VI

Bibliography

The following publications are referred to in the text.

Bishop, A. W., *The Strength of Soils as Engineering Materials*, Geotechnique, 1966, **16**, No. 2.

Clough, R. W., *The Finite Element Method in Structural Mechanics, Stress Analysis*, John Wiley & Sons Ltd., Chichester, 1965.

Clough, R. W., and Woodward, R. J., *Analysis of Embankment Stresses and Deformations*, Proc. Am. Soc. CE, SM and F Div., 1968.

Girijavallabhan, C. V., and Reese, L. C., *Finite Element Method for Problems in Soil Mechanics*, Proc. Am. Soc. CE, SM and F Div., 1968.

National Physical Laboratory, *Modern Computer Methods*, HMSO, 1961.

Wilson, E. L., *Finite Element Analysis of Two Dimensional Structures*, Structural Eng. Lab. Report No. 63, California University (Berkeley) 1963.

Zienkiewicz, O. C., *The Finite Element Method in Structural and Continuum Mechanics*, McGraw-Hill Book Co. Inc, New York, 1967.

Index